Biomedical Library

Queen's University Belfast

Tel: 028 9097 2710

E-mail: biomed.info@qub.ac.uk

For due dates and renewals:

QUB borrowers see 'MY ACCOUNT' at

http://library.qub.ac.uk/qcat

or go to the Library Home Page

HPSS borrowers see 'MY ACCOUNT' at

www.honni.qub.ac.uk/qcat

This book must be returned not later than its due
date, but is subject to recall if in demand

Fines are imposed on overdue books

OXFORD MEDICAL PUBLICATIONS

Nicotine, Smoking, and
The Low Tar Programme

This publication represents the proceedings of a symposium 'Nicotine, smoking, and the low tar programme' held on 18–20 November 1986, jointly at the Medical Research Council, 20 Park Crescent, London, W1N 4AL and Ciba Foundation, 41 Portland Place, London, W1N 4BN, and sponsored by the Tobacco Products Research Trust, Keats House, Guy's Hospital, St Thomas Street, London, SE1 9RT.

NICOTINE, SMOKING, AND THE LOW TAR PROGRAMME

EDITED BY

NICHOLAS WALD

and

SIR PETER FROGGATT

St Bartholomew's Hospital Medical College

OXFORD   NEW YORK   TOKYO
OXFORD UNIVERSITY PRESS
1989

Oxford University Press, Walton Street, Oxford OX2 6DP
Oxford New York Toronto
Delhi Bombay Calcutta Madras Karachi
Petaling Jaya Singapore Hong Kong Tokyo
Nairobi Dar es Salaam Cape Town
Melbourne Auckland
and associated companies in
Berlin Ibadan

Oxford is a trade mark of Oxford University Press

Published in the United States
by Oxford University Press, New York

© Nicholas Wald and Sir Peter Froggatt, 1989

British Library Cataloguing in Publication Data
Nicotine, smoking, and the low tar programme.
1. Tobacco smoking. Health aspects
I. Wald, Nicholas II. Froggatt, Sir Peter
613.8'5
ISBN 0-19-261729-X

Library of Congress Cataloging in Publication Data
Nicotine, smoking, and the low tar programme / edited by Nicholas Wald
and Sir Peter Froggatt.
p. cm. — (Oxford medical publications)
Proceedings of a symposium held in London on Nov. 18–20, 1986,
organized by the Independent Scientific Committee on Smoking and Health.
Includes bibliographies and index.
1. Tobacco—Toxicology—Congresses. 2. Nicotine—Toxicology—
Congresses. 3. Nicotine—Health aspects—Congresses. 4. Tar—
Health aspects—Congresses. 5. Smoking—Great Britain—Congresses.
I. Wald, Nicholas J. II. Froggatt, Peter. III. Great Britain.
Independent Scientific Committee on Smoking and Health. IV. Series.
[DNLM: 1. Nicotine—toxicity—congresses. 2. Smoking—adverse
effects—congresses. 3. Smoking—mortality—Great Britain—
congresses. 4. Smoking—trends—congresses. 5. Tars—congresses.
QV 137 N6624 1986] RA 1242.T6N53 1989 616.86'5—dc19 DNLM/DLC 88–23884
ISBN 0-19-261729-X

Set by Footnote Graphics, Warminster, Wiltshire
Printed in Great Britain by
Biddles Ltd., Guildford & King's Lynn

Preface

This Symposium, held on 18–20 November 1986, was organized by the Independent Scientific Committee on Smoking and Health and funded by the Tobacco Products Research Trust. It is the first of a series of symposia to be held under the auspices of the Committee which aim at drawing together important information which might have a bearing on future smoking and health policies.

The Committee was established in 1973. Its terms of reference are to advise the government on the health aspects of smoking and in some instances also to advise the tobacco industry. Its advice has always been that smoking is harmful to health, that smokers should stop smoking and non-smokers should not start, and that no level of smoking is free of risk. The Committee recognizes that many will not take this advice and accepts that there is a public duty to ensure that the toxicity of cigarettes is systematically reduced. It discharges this mainly through recommending that the tar yields of cigarettes are reduced and that the yields are regularly monitored not only with respect to tar, but also for other smoke components that may affect the toxicity of cigarettes. The Committee also assesses the acceptability of additives to tobacco products and offers advice on the scientific merit of research proposals submitted to the Tobacco Products Research Trust for financial support.

The Committee's *Third Report* (1984) drew attention to the possible toxicity of nicotine and its role in initiating and maintaining the smoking habit. By 1986 these questions had become pressing and for three reasons. First, evidence had accumulated that nicotine confers certain benefits on the smoker; this needed scrutiny. Secondly, nicotine was likely to play an important role in 'compensatory' smoking, which in this context is taken to be the process by which smoke from a cigarette with a high nicotine yield and a given tar yield is inhaled to a lesser extent than the smoke from a cigarette yielding less nicotine but the same tar yield, thereby reducing the intake of tar. The Committee's *Third Report* (para. 20) stated:

We also believe that there should be available to the public some brands with tar yields below those of the present principal Low Tar brands (i.e. below about 8 mg/ cigarette), but with proportionately higher nicotine yields (up to about 1 mg). There should be careful monitoring of public acceptance of such brands, and of the extent of 'compensation' both in those who use them and in those who smoke more conventional low tar cigarettes.

In line with the general thrust of this recommendation nicotine yields have declined less than tar yields. They are now relatively static, and it is therefore timely to assess to what extent nicotine can be said to determine compensation.

Thirdly, urgent consideration needed to be given to whether or not nicotine may be a co-carcinogen and its maintenance in cigarettes at current levels therefore ill-advised. This was perhaps the most important reason for holding the Symposium and publishing these Proceedings. Other aspects on the toxicity of nicotine were also considered and one, namely the possibility that nicotine may affect hormone-dependent disorders such as endometrial cancer and osteoporosis, was selected as the topic for the next symposium in the series. By that time results and ideas emerging from recent work should be available.

Any scientific judgement on the public health aspects of nicotine, smoking, and the low tar programme will necessarily be incomplete since new knowledge will always justify a reconsideration of the evidence. We hope that these Proceedings of the Symposium will form a base on which informed public health decisions can be made and, if necessary, revised in the light of new evidence.

We would like to thank those who participated in the Symposium, the members and scientific secretariat of the Independent Scientific Committee on Smoking and Health, the Trustees of the Tobacco Products Research Trust for their support, and Stephanie Kiryluk for her help in providing us with information used in these Proceedings. We are particularly indebted to Mrs Cheryl Swann for her assistance in organizing the meeting and the preparation of these Proceedings.

St Bartholomew's Hospital Medical College   N. W.
London   P. F.
February 1988

Contents

Contributors

Franz Adlkofer
Forschungsrat Rauchen u Gesundheit, Harvestehuder Weg 88, D-2000 Hamburg 13, Federal Republic of Germany.

Neal Benowitz
San Francisco General Hospital and Medical Center, Building 30, 5th Floor, 1001 Potrero Avenue, San Francisco, California 94110, USA.

A. Biber
Forschungsrat Rauchen u Gesundheit, Harvestehuder Weg 88, D-2000 Hamburg 13, Federal Republic of Germany.

Ian Cambell
Sully Hospital, Penarth, S. Glamorgan, CF6 2YA, UK.

Tim Clark
Dean, Guy's Hospital Medical School, London SE1 9RT, UK.

Sarah Darby
Imperial Cancer Research Fund, Cancer Epidemiology and Clinical Trials Unit, Gibson Laboratories, Radcliffe Infirmary, Oxford OX2 6HE, UK.

Richard Doll
Imperial Cancer Research Fund, Cancer Epidemiology and Clinical Trials Unit, Gibson Laboratories, Radcliffe Infirmary, Oxford OX2 6HE, UK.

Peter Elmes
Dawros House, St Andrews Road, Dinas Powys, S. Glamorgan CF6 4HB, UK.

Frank Fairweather
Unilever, Unilever House, PO Box 68, London EC4 4BQ, UK.

Sir Peter Froggatt
Department of Environmental and Preventive Medicine, St Bartholomew's Hospital Medical College, Charterhouse Square, London EC1M 6BQ, UK.

Christopher Frost

Department of Environmental and Preventive Medicine, St Bartholomew's Hospital Medical College, Charterhouse Square, London EC1M 6BQ, UK.

W.-D. Heller

Institut für Statistik und Mathematische Wirtschaftstheorie der Universität Karlsruhe, D-7500 Karlsruhe, Federal Republic of Germany.

Dietrich Hoffmann

American Health Foundation, Naylor Dana Institute for Disease Prevention, Dana Road, Valhalla, New York 10595, USA.

Ilse Hoffmann

American Health Foundation, Naylor Dana Institute for Disease Prevention, Dana Road, Valhalla, New York 10595, USA.

Walter Holland

Department of Community Medicine, St Thomas's Hospital Medical School, London SE1 7EH, UK.

Martin Jarvis

Addiction Research Unit, Institute of Psychiatry, 101 Denmark Hill, London SE5 8AF, UK.

Stephanie Kiryluk

Department of Environmental and Preventive Medicine, St Bartholomew's Hospital Medical College, Charterhouse Square, London EC1M 6BQ, UK.

E. Lawless

Labstat Inc., Kitchener, Ontario, Canada.

Peter N. Lee

Statistics and Computing Ltd, 17 Cedar Road, Sutton, Surrey SM2 5DA, UK.

Leonard Levy

Institute of Occupational Hygiene, University of Birmingham, Birmingham B15 2TJ, UK.

Alan Marsh

Social Survey Division, OPCS, London WC2, UK.

Patricia A. Martin

Institute of Occupational Hygiene, University of Birmingham, Birmingham B15 2TJ, UK.

Jil Matheson
Social Survey Division, OPCS, London WC2, UK.

Stuart Pocock
Department of Clinical Epidemiology and General Practice, Royal Free Hospital Medical School, Rowland Hill Street, London NW3 2QG, UK.

Ruth Porter
Medical Research Council, 20 Park Crescent, London W1N 4AL, UK.

William Rickert
Department of Statistics and Actuarial Science, University of Waterloo, Waterloo, Ontario N2L 3GI, Canada.

J. C. Robinson
Department of Statistics and Actuarial Science, University of Waterloo, Waterloo, Ontario N2L 3GI, Canada.

Francis Roe
19 Marryat Road, Wimbledon Common, London SW19 5BB, UK.

Michael Russell
Addiction Research Unit, Institute of Psychiatry, 101 Denmark Hill, London SE5 8AF, UK.

G. Scherer
Forschungsrat Rauchen u Gesundheit, Harvestehuder Weg 88, D-2000 Hamburg 13, Federal Republic of Germany.

Helmut Schievelbein
Holzbachstrasse 10, D-8034 Germering, Federal Republic of Germany.

Alison Stephen
Department of Environmental and Preventive Medicine, St Bartholomew's Hospital Medical College, Charterhouse Square, London EC1M 6BQ, UK.

Irene Stratton
Imperial Cancer Research Fund, Cancer Epidemiology and Clinical Trials Unit, Gibson Laboratories, Radcliffe Infirmary, Oxford OX2 6HE, UK.

Simon Thompson
Department of Environmental and Preventive Medicine, St Bartholomew's Hospital Medical College, Charterhouse Square, London EC1M 6BQ, UK.

Hugh Tunstall Pedoe

Cardiovascular Epidemiology Unit, Ninewells Hospital and Medical School, Dundee DD1 9SY, UK.

Nicholas Wald

Department of Environmental and Preventive Medicine, St Bartholomew's Hospital Medical College, Charterhouse Square, London EC1M 6BQ, UK.

David Warburton

Department of Psychology, University of Reading, Earley Gate, Whiteknights, Reading RG6 2AL, UK.

I

The Toxicity of Nicotine

1

Central nervous system toxicity of nicotine

NEAL L. BENOWITZ

Abstract

Consideration of a governmental smoking policy which might encour-
age the marketing of cigarettes delivering an increased ratio of nicotine
to tar must address the issue of potential toxicity of nicotine per se. The
evidence for toxic actions of nicotine are considered and discussed
together with changes in central nervous system (CNS) function result-
ing from prolonged exposure to nicotine as well as diseases of the CNS
influenced by cigarette smoking and with a suspected role of nicotine.
There is no evidence that nicotine produces any direct, permanent
CNS toxicity; it may contribute to the increased risk of stroke in
cigarette smokers and delay or protect against the development of
Parkinsonism.

1.1. Actions of nicotine on the central nervous system

A brief review of the actions of nicotine on the central nervous system is
necessary to appreciate any potential toxicity. Nicotine is a tertiary amine
composed of a pyridine and pyrrolidine ring. It binds to acetylcholine
receptors at ganglia, neuromuscular junctions, and in the brain. It readily
crossed the blood – brain barrier and is distributed throughout the
brain.[1-3] Uptake into the brain appears to involve both passive diffusion
and active transport by the choroid plexus.[4] Within the brain, specific
binding of tritiated nicotine has been shown to be greatest in the hypo-
thalamus, hippocampus, thalamus, midbrain, and brainstem, as well as
areas of the cerebral cortex.[5,6] Tritiated nicotine is also bound in the
nigrostriatal and mesolimbic dopaminergic neurons;[7] this may be important in understanding interactions between cigarette smoking and Parkinsonism, as will be discussed later. Autoradiographic studies of 2-
deoxyglucose uptake following nicotine administration show increased
metabolic activity in areas which correspond to areas of high degrees of
nicotine binding.[8]

In addition to, or as a consequence of, stimulation of nicotinic receptors,
acute exposure to nicotine results in activation of several central nervous

system (CNS) neurohumoral systems, with norepinephrine, dopamine, serotonin, vasopressin, growth hormone, and ACTH release.[9,10]

The physiological and behavioural effects of nicotine in humans are summarized in Table 1.1. Its cognitive actions are discussed in detail by Warburton (Chapter 13). These effects can occur by actions on the central nervous system itself, either directly or via peripheral chemoreceptors, by actions on the peripheral nervous system, via ganglionic stimulation, release of catecholamines from neurons and/or facilitation of neurotransmitter release, or on the adrenal medulla where it causes release of epinephrine.

The multiple sites of action and perhaps different receptor populations may explain unusual dose response relationships for nicotine. For some effects, low doses produce stimulation while higher doses produce relaxation or sedation.[11,12]

1.2. Adaptation to neural effects of nicotine

Tolerance develops rapidly to many of the effects of nicotine including dizziness, nausea, and subjective stimulant actions,[13] its extent depending

---

Table 1.1. Actions of nicotine in man

Cardiovascular
  Increased heart rate
    Cardiac contractility
    Blood pressure
  Cutaneous vasoconstriction—decreased skin temperature
  Catecholamine release

Metabolic
  Increased free fatty acids
    Glycerol
    Lactate

Central nervous system
  Arousal or relaxation
  EEG changes
  Tremor

Endocrine
  Increased growth hormone
    ACTH/cortisol
    Vasopressin
    β-endorphins
  Inhibition of prostacyclin synthesis

---

on the level of nicotine, the duration of exposure to a given level, and the rate of increase of nicotine level within a particular organ. Rapid dosing resulting in high brain concentrations of nicotine, as occurs with cigarette smoking, is presumably able to overcome some degree of tolerance, which is why people continue to smoke throughout the day. The time course of disappearance of tolerance may be an important determinant of individual cigarette smoking patterns. After overnight abstinence from cigarette smoke, people regain sensitivity to some effects of nicotine, such as arousal or heart rate acceleration, but not usually to other effects such as dizziness and nausea.

Changes in binding of nicotine in the brains of rats have been reported during prolonged dosing with nicotine.[14] Associated with development of behavioural tolerance, increased binding of nicotine was observed. The reasons for increased binding (rather than decreased binding as is seen with receptor down-regulation such as with β-adrenergic and other receptors) are as yet unknown, but are speculated to be related to receptor inactivation or generation of receptor antagonists (possibly a nicotine metabolite) during chronic nicotine treatment. Not surprisingly, with chronic exposure to nicotine, central neurotransmitter activity also changes, with decreased (in contrast to the initially increased) norepinephrine and dopamine turnover (personal communication, Darrell G. Kirch).

Abstinence from nicotine (derived from cigarettes, smokeless tobacco, or nicotine gum) results in a syndrome which includes irritability, dysphoria, lethargy, anxiety, impaired ability to concentrate, impaired psychomotor performance, and weight gain.[15–17]

From the above, it is clear that chronic nicotine exposure is associated with substantial central nervous system adaptation, including changes in receptors and neurotransmitter systems. A consequence of this adaptation is the development of abstinence symptoms during nicotine withdrawal. Whether there are long-term changes in central nervous system function as a consequence of chronic nicotine exposure is unknown.

1.3. Nicotine and cerebral circulation

Although not a neurological effect *per se*, nicotine may affect brain function by effects on the cerebral circulation. Chronic cigarette smoking accelerates atherosclerosis of coronary, peripheral and cerebral blood vessels. Nicotine may contribute to atherogenesis by effects on blood lipids or coagulation. Cigarette smokers have elevated low density (LDL) and very low density lipoproteins (VLDL), and reduced high density lipoprotein (HDL) concentrations compared to non-smokers, a profile associated with an increased risk of atherosclerosis.[18] It is hypothesized that nicotine, by releasing free fatty acids, increases the synthesis of triglycerides and VLDL by the liver, which in turn results in decreased HDL production.

The blood of smokers is known to coagulate more easily. Platelets in many studies are more reactive and platelet survival is shortened in smokers compared to non-smokers.[19,20] Thrombosis is believed to play an important role in atherogenesis.[21] Platelets may release a growth hormone which promotes the growth of vascular endothelial cells, contributing to the atherosclerotic plaque. The importance of nicotine as a determinant of platelet hyper-reactivity is supported by a study showing an apparent relationship between concentrations of nicotine after smoking different cigarettes and platelet aggregation response.[22] Nicotine could affect platelets by releasing epinephrine, which is known to enhance platelet reactivity, by inhibiting prostacyclin synthesis, an antiaggregatory hormone secreted by endothelial cells, or perhaps directly. Finally, by increasing heart rate and cardiac output, nicotine might increase blood turbulence and promote endothelial injury. Although several potential mechanisms for promoting atherogenesis have been considered, nicotine has not been demonstrated to accelerate atherosclerosis in experimental animals.

Chronic cigarette smoking is associated with decreased cerebral blood flow, which reverses within a year of stopping smoking.[23-25] Nicotine could contribute to reduced cerebral blood flow by promoting atherosclerosis, as discussed above, or by direct effects on cerebral blood vessels. Although nicotine causes vasoconstriction in some vascular beds, in humans and healthy animals it increases or does not change cerebral blood flow.[26-28] However, chronic cigarette smoking also impairs cerebral vasomotor reactivity, which could impair cerebral autoregulation.[29] Effects of nicotine on cerebral blood flow in the presence of cerebrovascular disease have not been investigated.

1.4.   Cigarette smoking, nicotine, and central nervous system disease

Stroke

Cigarette smoking has been known for many years to be a risk factor for development of subarachnoid haemorrhage.[30] Recently, several studies have shown that cigarette smoking is an independent risk factor for development of strokes of all types.[31-33] The increased risk of developing a stroke declines once an individual stops smoking.[32] Reduced cerebral blood flow is reported to be present in people prior to their development of stroke.[23] As discussed above, cigarette smoking is associated with reduction of cerebral blood flow which is, to some degree, reversible. It is conceivable that nicotine contributes to stroke by enhancing atherosclerosis, promoting thrombosis or by impairing cerebral autoregulation.

Nicotine could also contribute to subarachnoid haemorrhage by constricting or promoting thrombosis of nutrient arterioles which supply the large cerebral arteries, resulting in injury and weakness of the arterial wall and eventual rupture. Thus, although the role of nicotine in causing stroke has not been established, based on what is known of the pathogenesis of stroke and the pharmacological actions of nicotine, there is some basis for concern.

Parkinsonism

Four prospective and 12 case controlled studies have reported a negative association between cigarette smoking and Parkinsonism.[34] The relative risk averages about 0.5, indicating a protective effect. This association cannot be explained on the basis of selective early mortality in cigarette smokers or the use of control groups with smoking associated diseases.

Parkinsonism is characterized by depletion of dopamine in the basal ganglia. As noted previously, nicotine binds to dopaminergic neurons in this area of the brain and can increase dopamine turnover in the striatum.[7,35,36] Acute growth hormone release, which has been observed with cigarette smoking, is consistent with a central release of dopamine. Administration of nicotine has been reported to decrease the tremor in some Parkinsonian patients.[37] On this basis, it is suspected that the action of nicotine on the dopaminergic system of the basal ganglia delays or prevents the development of Parkinsonism. Thus, with respect to this disease, there may be a beneficial rather than a toxic effect of greater nicotine exposure.

1.5. Conclusions

While there are clearly significant changes in CNS receptors and the activity of neurotransmitter systems during prolonged exposure to nicotine, there is no evidence that nicotine produces any direct, permanent CNS toxicity.

Although there is no direct evidence of toxicity, there is concern, based on the known pathophysiology of the disease and the pharmacological actions of nicotine, that nicotine may contribute to the increased risk of stroke in cigarette smokers. Pathophysiological and pharmacological considerations suggest that nicotine in cigarette smokers delays or protects against the development of Parkinsonism.

Even if we accept an aetiological association between nicotine exposure and stroke and a protective effect of nicotine against Parkinsonism, no dose–response data are available with which to estimate the risks or benefits attributable to a particular level of exposure to nicotine. This is important because for other effects, such as heart rate acceleration, the dose–response relationship for nicotine is not linear and lower exposure levels have similar cardiovascular effects to higher levels.[38] If such dose–response relationships hold for CNS actions, then changes in nicotine intake might not influence health effects.

The benefit of reducing the cancer risk by reducing tar exposure is likely to be substantial. If tar exposure can be reduced by increasing the nicotine-to-tar yield ratio of cigarettes, any concern about the CNS toxicity of nicotine is not sufficient to argue against implementation of such a policy.

Acknowledgements

Research supported in part by grants DA02277, CA32389, and DA01696 from the National Institutes of Health.

References

1. Oldendorf, W. H. (1974). Lipid solubility and drug penetration in the blood–brain barrier. *Proceedings of the Society for Experimental and Biological Medicine*, **147**, 813–6.
2. Stalhandski, T. (1970). Effects of increased liver metabolism of nicotine on its uptake, elimination and toxicity in mice. *Acta Physiologica Scandinavica*, **80**, 222–34.
3. Maziere, M., Comar, D., Marazano, C., and Berger, G. (1976). Nicotine-[11]C: Synthesis and distribution kinetics in animals. *European Journal of Nuclear Medicine*, **1**, 255–8.
4. Spector, R. and Goldberg, M. J. (1982). Active transport of nicotine by the isolated choroid plexus *in vitro*. *Journal of Neurochemistry*, **38**, 594–6.
5. Clarke, P. B. S., Pert, C. B., and Pert, A. (1984). Autoradiographic distribution of nicotine receptors in rat brain. *Brain Research*, **323**, 390–5.
6. London, E. D., Waller, S. B., and Wamsley, J. K. (1985). Autoradiographic localization of (^{3}H)nicotine binding sites in the rat brain. *Neuroscience Letters*, **53**, 179–84.
7. Clarke, P. B. S. and Pert, A. (1985). Autoradiographic evidence for nicotine receptors on nigrostriatal and mesolimbic dopaminergic neurons. *Brain Research*, **348**, 355–8.
8. London, E. D., Connolly, R. J., Szikszay, M., and Wamsley, J. K. (1985). Distribution of cerebral metabolic effects of nicotine in the rat. *European Journal of Pharmacology*, **110**, 391–2.
9. Balfour, D. J. K. (1982). The effects of nicotine on brain neurotransmitter systems. *Pharmacology and Therapeutics*, **16**, 269–82.
10. Pomerleau, O. F. and Pomerleau, C. S. (1984). Neuroregulators and the

reinforcement of smoking: Towards a biobehavioral explanation. *Neuroscience and Biobehavioral Reviews*, **8**, 503–13.

11. Armitage, A. K., Hall, G. H., and Morrison, C. F. (1968). Pharmacological basis for the tobacco smoking habit. *Nature*, **217**, 331–4.

12. Ashton, H., Marsh, V. R., Millman, J. E., Rawlins, M. D., Telford, R., and Thompson, J. W. (1980). Biphasic dose-related responses of the CNV (contingent negative variation) to I.V. nicotine in man. *British Journal of Clinical Pharmacology*, **10**, 579–89.

13. Jones, R. T., Farrell, T. R., and Herning, R. I. (1978). Tobacco smoking and nicotine tolerance. In *Self-administration of abused substances: methods for study*, National Institute on Drug Abuse Research Monograph Series, No. 18 (ed. N. A. Krasnegor), pp. 202–8. US Government Printing Office, Washington, DC.

14. Marks, M. J., Stitzel, J. A., and Collins, A. C. (1985). Time course study of the effects of chronic nicotine infusion on drug response and brain receptors. *Journal of Pharmacology and Experimental Therapeutics*, **235**, 619–28.

15. Hatsukami, D. K., Hughes, J. R., Pickens, R. W., and Svikis, D. (1984). Tobacco withdrawal symptoms: An experimental analysis. *Psychopharmacology*, **84**, 231–6.

16. West, R. J. and Russell, M. A. H. (1985). Effects of withdrawal from long-term nicotine gum use. *Psychological Medicine*, **15**, 391–3.

17. Hatsukami, D. K., Gust, S. W., and Keenan, R. M. (1987). Physiological and subjective changes from smokeless tobacco withdrawal. *Clinical Pharmacology and Therapeutics* **41**, 103–7.

18. Brischetto, C. S., Connor, W. E., Connor, S. L., and Matarazzo, J. D. (1983). Plasma lipid and lipoprotein profiles of cigarette smokers from randomly selected families: Enhancement of hyperlipidemia and depression of high-density lipoprotein. *American Journal of Cardiology*, **52**, 675–80.

19. Billimoria, J. D., Pozner, H., Metselaar, B., Best, F. W., and James, D. C. O. (1975). Effect of cigarette smoking on lipids, lipoproteins, blood coagulation, fibrinolysis and cellular components of human blood. *Atherosclerosis*, **21**, 61–76.

20. Mustard, J. F. and Murphy, E. A. (1963). Effect of smoking on blood coagulation and platelet survival in man. *British Medical Journal*, **1**, 846–9.

21. Mehta, J. and Mehta, P. (1981). Role of blood platelets and prostaglandins in coronary artery disease. *American Journal of Cardiology*, **48**, 366–73.

22. Renaud, S., Blache, D., Dumont, E., Thevenon, C., and Wissendanger, T. (1984). Platelet function after cigarette smoking in relation to nicotine and carbon monoxide. *Clinical Pharmacology and Therapeutics*, **36**, 389–95.

23. Rogers, R. L., Meyer, J. S., Shaw, T. G., Mortel, K. F., Hardenberg, J. P., and Zaid, R. R. (1983). Cigarette smoking decreases cerebral blood flow suggesting increased risk for stroke. *Journal of the American Medical Association*, **250**, 2796–800.

24. Kubota, K., Yamaguchi, T., Abe, Y., Fujiwara, T., Hatazawa, J., and Matsuzawa, T. (1983). Effects of smoking on regional cerebral blood flow in neurologically normal subjects. *Stroke*, **14**, 720–4.

25. Rogers, R. L., Meyer, J. S., Judd, B. W., and Mortel, K. F. (1985). Abstention from cigarette smoking improves cerebral perfusion among elderly chronic smokers. *Journal of the American Medical Association*, **253**, 2970–4.

26. Miyazaki, M. (1969). Circulatory effect of cigarette smoking, with special reference to the effect on cerebral hemodynamics. *Japanese Circulation Journal*, **33**, 907–12.
27. Skinhoj, E., Olesen, J., and Paulson, O. B. (1973). Influence of smoking and nicotine on cerebral blood flow and metabolic rate of oxygen in man. *Journal of Applied Physiology*, **35**, 820–2.
28. Crystal, G. J., Downey, H. F., Adkins, T. P., and Bashour, F. A. (1983). Regional blood flow in canine brain during nicotine infusion: Effect of autonomic blocking drugs. *Stroke*, **14**, 941–7.
29. Rogers, R. L., Meyer, J. S., Shaw, T. G., Mortel, K. F., and Thornby, J. (1984). The effects of chronic cigarette smoking on cerebrovascular responsiveness to 5 per cent CO_2 and 100 per cent O_2 inhalation. *Journal of the American Geriatrics Society*, **32**, 415–20.
30. Sacco, R. L., *et al.* (1984). Subarachnoid and intracerebral haemorrhage: Natural history, prognosis and precursive factors in the Framingham Study. *Neurology*, **34**, 847–54.
31. Khaw, K-T., Barrett-Connor, E., Suarez, L., and Criqui, M. H. (1984). Predictors of stroke-associated mortality in the elderly. *Stroke*, **15**, 244–8.
32. Abbott, R. D., Yin, Y., Reed, D. M., and Yano, K. (1986). Risk of stroke in male cigarette smokers. *New England Journal of Medicine*, **315**, 717–20.
33. Bonita, R., Scragg, R., Stewart, A., Jackson, R., and Beaglehole, R. (1986). Cigarette smoking and risk of premature stroke in men and women. *British Medical Journal*, **293**, 6–8.
34. Baron, J. A. (1986). Cigarette smoking and Parkinson's disease. *Neurology*, **36**, 1490–4.
35. Clarke, P. B. S., Hommer, D. W., Pert, A., and Skirboll, L. R. (1985). Electrophysiological actions of nicotine on substantia nigra single units. *British Journal of Pharmacology*, **85**, 828–35.
36. Giorguieff-Chesselet, M. F., Kemel, M. L., Wandscheer, D., and Glowinski, J. (1979). Regulation of dopamine release by presynaptic nicotinic receptors in rat striatal slices; Effect of nicotine in a low concentration. *Life Sciences*, **25**, 1257–62.
37. Marshall, J. and Schnieden, H. (1966). Effect of adrenaline, noradrenaline, atropine, and nicotine on some types of human tremor. *Journal of Neurology Neurosurgery and Psychiatry*, **29**, 214–3.
38. Benowitz, N. L., Kuyt, F., and Jacob, P. (1984). Influence of nicotine on cardiovascular and hormonal effects of cigarette smoking. *Clinical Pharmacology and Therapeutics*, **36**, 74–81.

2

Toxicology of nicotine – its role in the aetiology of cancer due to cigarette smoking and cardiovascular disease

L. S. LEVY and P. A. MARTIN

Abstract

The introduction of the low tar programme, which involves consequentially lower nicotine levels, was intended to reduce the rate of lung cancer in cigarette smokers. The programme assumed that nicotine per se had no role in the pathogenesis of lung cancer, an assumption now in question. If it has a role then a further reduction could be warranted and the recommendation by the Third Report of the ISCSH and others, viz., that reducing the tar/nicotine ratio can play an important and hazard-free part in reducing the sales-weighted average tar yield, and, therefore the risk of lung cancer in committed smokers, would be unwise. On the other hand a further reduction could lead to smoking 'compensation' and an increase in tar absorption which could have a net result of increased lung cancer risk whether or not nicotine plays any role in carcinogenicity. In this situation the smoking 'compensation' is achieved as a result of an adaptation of smoking habit, whereby a smoker self-titrates body nicotine levels to suit his/her own pharmacological dependence. This paper introduces some of the arguments and evidence in this area and briefly reviews the possible role of nicotine in cardiovascular disease.

2.1. Introduction

Nicotine, 1-methyl-2-(3-pyridyl)pyrrolidine, is a naturally occurring alkaloid present in *Nicotiana tabacum* and is largely responsible for the pharmacological effects of cigarette smoke. The alkaloid can be extracted from the leaves and stems of tobacco plants where it is present in concentrations of 2–14 per cent. It is a volatile, strongly alkaline liquid which develops a pyridine or tobacco-like odour on exposure to air. Cigarette smoke itself contains at least ten or more major alkaloids of which nicotine is one example (Fig. 2.1). It has been estimated that the average cigarette

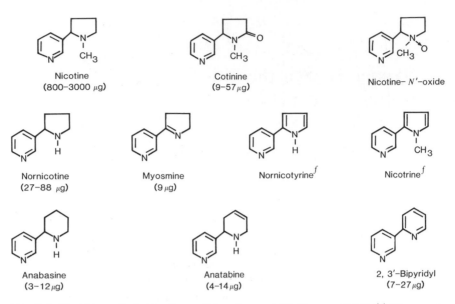

Fig. 2.1. Pyridine alkaloids (from Schmeltz and Hoffmann 1977).[A4] Figures in parentheses indicate levels in the mainstream smoke per cigarette. *f*, Quantitative data not available.

smoker who inhales absorbs, with every puff, a dose of nicotine which is equivalent to 0.1 mg nicotine given intravenously.[1]

Smoke from pipe or cigar tobacco is *alkaline* whereas that from cigarettes is *acid*. This means that pipe and cigar smoke, containing nicotine, may be absorbed through the lining of the mouth and pharynx thereby obviating the need for pipe and cigar smokers to inhale in order to achieve their pharmacologically-dependent levels of nicotine, whereas the nicotine in cigarette smoke is chiefly absorbed after inhalation of the smoke via the lungs.

After uptake by the body, nicotine is metabolized to either cotinine or nicotine-N^1-oxide. Neither of these two substances appear to be pharmacologically active and both have been identified as components of cigarette smoke itself.

In addition to the nicotine found as a natural substance in the tobacco plant, modification of the parent compound often occurs in the curing and fermentation processes during manufacture of the end product. These processes lead to the formation of tobacco-specific nitrosamines (Fig. 2.2).[2] In addition, it has been demonstrated that these nitrosamines may be formed endogenously.

Cigarette smoking is the most pressing public health issue in economically developed countries and public health policy has been double-edged,

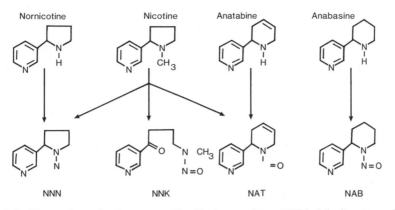

Fig. 2.2. Formation of tobacco-specific *N*-nitrosamines. NNN, *N'*-nitrosonornico-
tine; NNK, 4-(methylnitrosamino)-1-(3-pyridyl)-1-butanone; NAT, *N'*-nitroso-
anatabine; NAB, *N'*-nitrosoanabasine.

directed both at encouraging smokers to stop and discouraging non-
smokers from starting to smoke. To a certain extent these policies have
achieved limited success (Fig. 2.3).

By the early 1980's in both the US and the UK, cigarette consumption
per head had decreased. In the UK this decline was greater for males (35
per cent) than females (25 per cent). These figures were obtained from the
Tobacco Advisory Council (unpublished data). Trends in both countries
since the early 1970's have shown a movement by the consumer towards
smoking cigarettes with a *low tar and low nicotine yield* (Fig. 2.4).

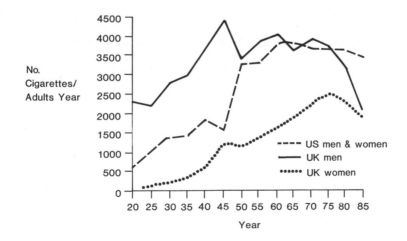

Fig. 2.3. Annual consumption of manufactured cigarettes per adult in the USA and
UK (1920–85). Data for men and women separately are not available in the USA.

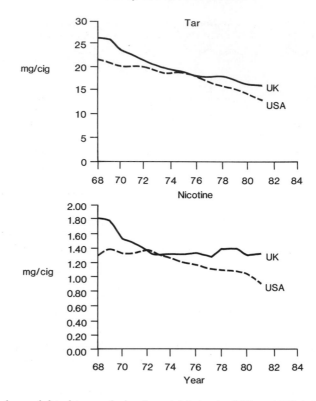

Fig. 2.4. Sales-weighted tar and nicotine yields in the UK and USA (1968–84).

This paper was presented primarily to set the scene for discussion on the possible aetiology of cancer due to nicotine. A more detailed account of the putative mechanism(s) by which nicotine may exert its carcinogenic effects is addressed by Hoffmann (Chapter 3).

2.2. Cancer and nicotine

Tobacco smoking, in the form of cigarettes, is the major single cause of cancer mortality and is responsible for the development of more tumours than all other known causes of cancer combined.[3] Table 2.1 shows the relative risk values for cancer at six sites experienced by smokers of cigarettes relative to non-smokers.

These increased cancer risks are not confined to cigarette smokers alone; both cigar and pipe smokers experience a doubling of the risk for lung cancer and a four-fold excess risk for cancer of the larynx. Although many of these risk estimates are based upon studies of individuals who smoked

Table 2.1. Relative risks of cancer in cigarette smokers

Cancer site	Relative risks	
	Range (from 8 cohort studies)	'Best' estimate
Lung	3.6–15.9	14.0
Larynx	6.1–13.6	8.0
Buccal cavity	1.0–13.0	4.0
Pharynx	2.8–12.5	4.0
Oesophagus	0.7– 6.6	3.0
Bladder	1.0– 6.0	2.0

cigarettes which were marketed in past decades, even the present day *low yield* cigarettes still present an appreciable increased risk of cancer development. Whilst the carcinogenic activity of tobacco smoke is apparently related to the tar content of the product,[4,A3] the cancer risk is *dose-related* to the smoke composition and, thus, it would appear reasonable to assume that manipulation of the product may influence this cancer risk. Thus, cigarettes yielding less tar/nicotine would present a reduced lung cancer risk. However, the reality of the situation is not so straightforward since both the tar and the nicotine content of tobacco products may be varied independently.

Manipulation of cigarette production has led to a reduction in both the tar and nicotine content, and with the introduction of filters, a concomitant reduction in exposure to certain polycyclic aromatic hydrocarbons and other particulate materials. These have been accepted by the consumer to a greater or lesser extent. However, the possible further reduction in nicotine levels in tobacco products is more controversial since it may influence the existing risks due to other components of the smoke as the consumer attempts to compensate for the reduced nicotine yield by altering his/her smoking method. Studies[5] have shown that these changes in smoking methods are related to the consumer's ability to *self-regulate* his/her own blood nicotine level. Higher nicotine concentrations may be obtained by altering the frequency and depth of inhalation,[6] or by mechanically compressing the filter tips or blocking the air channels in the filter.[7] As a result of these various practices, tar yields may be increased by 51 per cent, nicotine levels by 69 per cent, and carbon monoxide levels by 147 per cent over their published values.[8]

Thus, before any alteration of the nicotine content of cigarettes can be usefully considered, a number of key issues relating to the possible carcinogenicity of nicotine and nicotine-associated substances in tobacco must be addressed.

1. Is nicotine *per se* a carcinogen?

2. Are the products of metabolism within the body carcinogenic?

3. Are any nicotine-derived substances, produced during the curing or processing, or smoking of tobacco carcinogenic?

4. If nicotine, or any nicotine-derived substance, has an effect on human cancer, is this manifest by a direct mechanism, or by a non-genotoxic mechanism (promotion, etc.). If the latter applies, does this influence the activity of other known tobacco smoke carcinogens?

1. At the present time there is no sound evidence from animal studies which demonstrates that nicotine *per se* is a carcinogen. It should be noted, however, that the requirements of a contemporary protocol for a carcinogenesis bioassay would not be satisfied by those studies presently reported.

2. The principle stable urinary metabolites of nicotine, cotinine and nicotine-N^1-oxide (*cis-* and *trans-*), were examined for carcinogenic activity and the earlier reports of the induction of lymphomas by cotinine[9] were not confirmed by later studies using twice the dose level.[10] Later contemporary studies have shown that these metabolites are neither carcinogenic nor were they promoters of urinary bladder tumours when administered to F344 rats pretreated with *N*-[4-(5-nitro-2-furyl)-2-thiazolyl] formamide (FANFT) a recognized experimental bladder carcinogen.[11]
An interesting concept involves the metabolism of nicotine to cotinine via an unstable intermediate, the *iminium ion*. This nicotine-$\Delta 1'(5')$ iminium ion is able to form adducts with glutathione.[12] Thus, it would seem reasonable to propose that this unstable iminium ion may react with other cellular macromolecules and might be genotoxic. Further work in this area is required.

3. Tobacco-specific *N*-nitrosamines are formed during curing and processing by the nitrosation of the tertiary amine nicotine by nitrate, found mainly in the tobacco leaf stems and ribs. Cleavage of the N-CH_3 bond and the loss of formaldehyde yields NNN. Similar cleavage of either the 2'-N or 5'-N bond yields NNK or NNA, respectively (Fig. 2.2). Other tobacco alkaloids, such as anabasine, anatabine, and nornicotine may similarly react with nitrate to form NNN, NAB, and NAT. All of the tobacco-specific nitrosamines, except NNA, have been detected in cigarette, cigar, and snuff tobacco, and in mainstream and sidestream tobacco smoke[13] due to pyrolysis of the alkaloids with nitrogen oxides. In addition, there is some evidence that tobacco-specific nitrosamines are formed endogenously in smokers and snuff dippers.[4,13]

Experiments have shown that NNK, followed by NNN are the most potent carcinogens among the tobacco-specific nitrosamines. The site of tumour formation is dependent on the route of administration to some extent. NAT was inactive when tested in rats by subcutaneous injection.[2] Nitrosamines are enzymatically converted to unstable electrophilic intermediates which can react with nucleophilic centres in cellular macromolecules, such as DNA. Thus, the formation of NNK from nicotine during tobacco processing and cigarette smoking, and its subsequent metabolic conversion to a methylating agent provide a pathway by which the N-methyl group of nicotine can methylate DNA, a reaction which is considered crucial in causing an impairment and alteration of the structural integrity of the cell's heritable material. Hence, this clearly demonstrates a strong mechanistic link between nicotine and tobacco-related cancer.

Further supporting evidence for an association between nicotine-derived carcinogens and human cancer can be gained from a study of the high incidence of oral cancers among non-smoking, long-term snuff dippers. The causative agent(s) is presumed to be the tobacco-specific nitrosamines in this tobacco product since these chemicals represent the only known carcinogenic materials present at any appreciable level.

4. Studies by Bock and co-workers[14] in which nicotine was utilised in the mouse skin initiation-promotion model with either benzo(a)pyrene (BaP) or tobacco smoke condensate (TSC) failed to clearly demonstrate either promotion or co-carcinogenic properties. Nicotine may act to facilitate expression of other carcinogenic factors in tobacco smoke, but at this time evidence is not available to prove or disprove such speculation.

Conclusions

At the present time there is no experimental evidence that nicotine is carcinogenic *per se*. However, the studies which are available for evaluation are inadequate by contemporary standards. The principle urinary metabolites of nicotine are non-carcinogenic. However, with other pyridine alkaloids in the tobacco plant and together with nitrate (found in the stems and ribs) nicotine can be converted to a group of chemicals, the N-nitrosamines, many of which are potent animal carcinogens at a variety of sites and by various routes. Among all of the groups of known chemical carcinogens, the N-nitroso compounds possess characteristics which make them prime candidates as agents in the induction of human cancer,[15] despite the lack of good epidemiological data linking them with human cancer.

Thus in our opinion the statement, 'nicotine and its metabolites must be considered to play little or no role in carcinogenesis'[16] should be reconsidered. There is good scientific evidence for the carcinogenicity of a number of tobacco-specific nitrosamines (formed from nicotine) in animal studies. What is lacking at the present time is the link between these factors and human tobacco-related cancer. Moreover, it would be extremely difficult to design a study protocol which would attempt to causally relate these materials to cancer. Nevertheless, it would seem probable that low yield cigarettes would result in a reduction in the amount of nicotine available for nitrosation. This, in turn, might logically be expected to lead to a reduction in the potential for human lung cancer due to the action(s) of tobacco-specific nitrosamines. It would appear, therefore, that good circumstantial evidence might be the best upon which decision making can be based at the present time. Clearly, it would seem prudent to minimize exposure to the potential precursors for endogenous nitrosamine formation.

2.3.   Nicotine and cardiovascular effects

Smoking is generally accepted to be a major risk factor for coronary and peripheral vascular disease, although the mechanisms by which smoking contributes to the development of clinical manifestations of cardiovascular disease are not well understood. Although nicotine may play a role in the pathogenesis of these smoking-related diseases (Fig. 2.5) the evidence is circumstantial and not proven. Cigarette smoking causes the release of adrenaline and noradrenaline from the adrenal glands[1] and nicotine injection[17] has a similar effect. It is likely that the main actions of smoking on the heart and blood vessels are due to this hormonal release. In addition, nicotine also acts directly on the nervous centres, via the sympathetic nervous system, that control blood pressure, heart rate, and cardiac output, causing a transient increase in healthy people. There is a gradual return to baseline levels approximately 15 minutes after smoking a cigarette.[18,19]

In smokers with pre-existing coronary heart disease the effects are variable. There may be a reduction in cardiac output and ventricular contractility as well as vasoconstriction[A2,20] on smoking a cigarette. In addition, there may be cardiac arrhythmia. Nicotine tends to increase the concentration of fatty acids in the blood[21] and there is equivocal evidence to suggest that nicotine also increases the liability of blood platelets to adhere to each other and to the walls of blood vessels. These latter actions are observed during the early stages of *thrombosis* (clotting) of the blood, and may be linked to the formation of atheroma of the arteries. Further evidence for the role of nicotine in atheroma formation may be related to its known action in causing the release of epinephrine. This hormone is known to enhance platelet activity. An important additional risk factor for

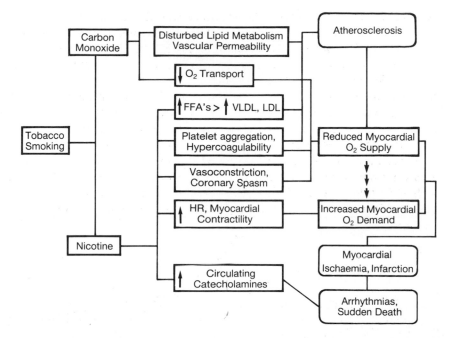

Fig. 2.5. Smoking, nicotine, and coronary heart disease (after Benowitz 1986).[27]

ischaemic heart disease (IHD) is clear evidence of an increase in plasma fibrinogen levels in smokers when compared to non-smokers.[22] A particularly interesting finding in this study by Meade and his colleagues was that cigarette smokers who change to cigars increased their fibrinogen levels even further. This correlates with the risk of IHD to cigar smokers. Any evaluation of the role of nicotine on cardiovascular events would be incomplete without consideration of the carbon monoxide component of cigarette smoke. The net result of the combined actions of nicotine and carbon monoxide on the cardiovascular system would be an increased demand for myocardial oxygen in the presence of a limited supply; a set of conditions that could potentially precipitate myocardial ischaemia.[23]

Smoking appears to be an independent risk factor in both sexes for most classes of cardiovascular disease, including coronary heart disease (CHD), peripheral vascular disease, and cerebrovascular disease. For CHD a dose–response relationship has been demonstrated, although its strength varies depending on the condition. Cigar and pipe smoking generally confer an increased risk of cardiovascular diseases, but the attributable risk is less than for cigarette smoking (Table 2.2).

Cessation of smoking reduces the risk of cardiovascular events and related mortality, with significant improvements observed in the first year

Table 2.2. Coronary heart disease mortality ratios for male cigarette, pipe, cigar, and mixed pipe and/or cigar smokers (cohort studies)

Study	Mortality ratios				
	Non-smoker	Cigarette smoker	Pipe smoker	Cigar smoker	Mixed smoker
US veterans	1.00	1.58	1.02	1.12	
American Cancer Society, 9-state	1.00	1.70	—	1.28	
Swedish	1.00	1.70	1.40		
American Cancer Society, 25-state	1.00	1.9–2.55	1.08		
British physicians	1.00	1.62			1.03

Data from Surgeon General report.[28]

of cessation, followed by a more gradual reduction in risk towards the non-smoker levels.

Whilst some studies suggest a significant reduction (20 per cent) in CHD mortality among low-tar cigarette smokers, others, including the Framingham study suggest that low-tar/nicotine and filter cigarette smokers do not experience a lower CHD incidence rate than those smoking high-tar, unfiltered brands.[24,A1] This finding has been corroborated in several case-control studies. Indeed, Kaufman *et al.*[25] reported that the relative risk of non-fatal myocardial infarction amongst young men (30–54 years old) smoking low-tar cigarettes was 2.8 times that of a non-smoker. Great care must be taken in the interpretation of the role of nicotine in all these findings, particularly when comparing relative risks for various smoking groups. As an example, the work of Wald and colleagues[26] has shown that the serum level of cotinine, the principal metabolite of nicotine, is highest in pipe smokers. This group have been shown to have the lowest risk of all smoking groups of CHD when compared to non-smokers (see Fig. 2.6). Many of these apparent anomalies are now thought to be due to the fact that either:

(1) smokers compensate for the reduced yields of nicotine and tar from their cigarettes by altering their smoking methods so as to maintain their optimum-required blood nicotine levels; or

(2) nicotine, as suggested by the work of Wald *et al.*[26] may not be the major determinant of cardiovascular disease.

Conclusions

There is evidence that nicotine is able to exert a variety of cardiovascular

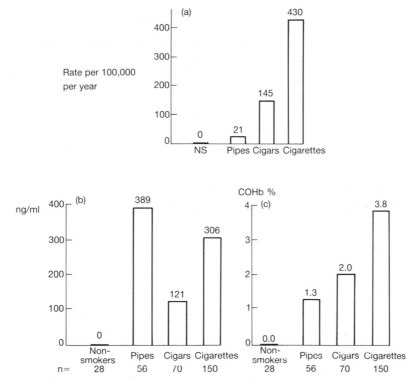

Fig. 2.6. (a) Excess coronary heart disease mortality. (b) Mean serum cotinine. (c) Mean excess COHb.

effects via stimulation of the nervous and endocrine systems. However, whilst there are plausible hypotheses which exist to implicate nicotine in the pathogenesis of cardiovascular disease, the epidemiological data available would seem to mitigate against nicotine being cardiotoxic. This reasoning is supported by the findings following the introduction of low-tar–low-nicotine cigarettes where a substantial reduction in the incidence of cardiovascular disease might have been expected, had nicotine been a major determinant. However, the results of a variety of epidemiological studies have been inconsistent in demonstrating such a reduction and further research is therefore needed to clarify the situation.

References

1. Armitage, A. (1965). Effects of nicotine and tobacco smoke on blood pressure and release of catecholamines from the adrenal glands. *British Journal of Pharmacology*, **25**, 515–26.
2. Hoffmann, D. and Hecht, S. S. (1985). Nicotine-derived N-nitrosamines and

tobacco-related cancer: current status and future directions. *Cancer Research*, **45**, 935–44.

3. Doll, R. and Peto, R. (1981). *The causes of cancer*. Oxford Medical Publications, Oxford.
4. Winn, D. M., Blot, W. J., Shy, C. M., Pickle, L. W., Toledo, A., and Fraumeni, J. F. (1981). Snuff dipping and oral cancer among women in the Southern United States. *New England Journal of Medicine*, **304**, 745–9.
5. Herning, R. I., Jones, R. T., Benowitz, N., and Mines, A. H. (1983). How a cigarette is smoked determines blood nicotine levels. *Clinical Pharmacology and Therapeutics*, **33**, 84–90.
6. Sutton, S. R., *et al.* (1982). Relationship between cigarette yields, puffing patterns and smoke intake: Evidence for tar compensation. *British Medical Journal*, **285**, 600–3.
7. Kozlowski, L. T., Frecker, R. C., Khouw, V., and Pope, M. A. (1980). The misuse of 'less hazardous' cigarettes and detection: Hole-blocking ventilated filters. *American Journal of Public Health*, **70**, 1202–3.
8. Hoffmann, D., Adams, J. D., and Haley, N. J. (1983). Reported cigarette smoke values: a closer look. *American Journal of Public Health*, **73**, 1050–3.
9. Truhaut, R., Declercq, M., and Loisillier, F. (1964). Acute and chronic toxicity of cotinine and its carcinogenic activity in the rat. *Pathologie Biologie*, **12**, 39–42.
10. Schmahl, D. and Osswald, H. (1968). Fehlen einer carcinogen wirkung von Cotinin bei ratten. *Zeitschrift fur Krebsforschung*, **71**, 198.
11. Lavoie, E. J., Shigematsu, A., Rivenson, A., Mu. B., & Hoffmann, D. (1985). Evaluation of the effects of cotinine and nicotine-N-oxides on the development of tumors in rats initiated with N-(4-(5-nitro-2-furyl)-2-thiazolyl)formamide. *Journal of the National Cancer Institute*, **75**, 1075–81.
12. Hibbard, A. R. and Gorrod, J. W. (1982). Nicotine-Δ1 (5′) iminium ion: a reactive intermediate in nicotine metabolism. *Advances in Experimental Medicine and Biology*, **136B**, 1121–31.
13. Hoffmann, D., Brunnemann, K. D., Adams, J. D., and Hecht, S. S. (1984). *Formation and analysis of N-nitrosamines in tobacco products and their endogenous formation in tobacco consumers*, *IARC Scientific Publications No. 57*, pp. 743–62. Lyon, France.
14. Bock, F. G. (1980). Cocarcinogenic properties of nicotine. In *Banbury Report 3 − A safe cigarette?* (ed. G. B. Gori and F. G. Bock), pp. 129–39. Cold Spring Harbor, New York.
15. Lijinsky, W. and Epstein, S. (1970). Nitrosamines as environmental carcinogens. *Nature*, **225**, 21.
16. Independent Scientific Committee on Smoking and Health (1983). *Third report of the ISCSH*. HMSO, London.
17. Larson, P. S. and Silvette, H. (1968). *Tobacco: experimental and clinical studies. A comprehensive account of world literature*. Williams & Wilkins Co., Baltimore.
18. Rosenberg, L., *et al.* (1983). Myocardial infarction in women under 50 years of age. *Journal of the American Medical Association*, **250**, 2801–6.
19. Klein, L. W. and Gorlin, R. (1983). The systemic and coronary hemodynamic response to cigarette smoking. *New York State Journal of Medicine*, **83**, 1264–5.

20. Pentecost, B. and Shillingford, J. (1964). The acute effects of smoking on myocardial performance in patients with coronary arterial disease. *British Heart Journal*, **26**, 422–9.
21. US Public Health Service (1968). The Health Consequences of Smoking. *1968 Supplement to the 1967 Public Health Service Review*. Publication No. 1696. Washington D.C.
22. Meade, T. W., Imeson, J., and Stirling, Y. (1987). Effects of changes in smoking and other characteristics on clotting factors and the risk of ischaemic heart disease. *Lancet*, **ii**, 986–8.
23. Kien, G. A. and Sherrod, T. (1960). Action of nicotine and smoking on coronary circulation and myocardial oxygen utilisation. *Annals of the New York Academy of Sciences*, **90**, 161–73.
24. Lee, P. N. and Garfinkel, L. (1981). Mortality and type of cigarette smoked. *Journal of Epidemiology and Community Health*, **35**, 16–22.
25. Kaufman, D. W., *et al.* (1983). Nicotine and carbon monoxide content of cigarette smoke and the risk of myocardial infarction in young men. *New England Journal of Medicine*, **308**, 407–13.
26. Wald, N. J., Idle, M., and Boreham, J. (1981). Serum cotinine levels in pipe smokers: evidence against nicotine as cause of coronary heart disease. *Lancet*, **ii**, 775–7.
27. Benowitz, N. L. (1986). Clinical pharmacology of nicotine. *Annual Review of Medicine*, **37**, 21–32.
28. Department Health & Human Services (1983). The health consequences of smoking: cardiovascular disease. In *A report of the Surgeon General*. Rockville, Maryland, USA USPHS.
A1. Castelli, W. P., Dawber, T. R., Feinlab, M., Garrison, R. J., McNamara, P. M., and Kannel, W. B. (1968). The filter cigarette and coronary heart disease: The Framingham study. *Lancet*, **ii**, 109–13.
A2. Frankl, W. S., Winters, W. L., and Soloff, L. A. (1965). The effects of smoking on the cardiac output at rest and during exercise in patients with healed myocardial infarction. *Circulation*, **31**, 42–4.
A3. Wynder, E. L. and Hoffmann, D. (1967). *Tobacco and tobacco smoke: Studies in experimental carcinogenesis*. NY Academic Press.
A4. Schmeltz, I. and Hoffmann, D. (1977). Nitrogen-containing compounds in tobacco and tobacco smoke, *Chemical Reviews*, **77**, 295–311.

3

Nicotine, a tobacco-specific precursor for carcinogens

DIETRICH HOFFMANN

Abstract

The evidence for the possible carcinogenicity of the nicotine metabolites cotinine and nicotine-N-oxides as well as nicotine per se is considered, but more importantly, the data for nicotine as a precursor for carcinogenic tobacco-specific nitrosamines (TSNA) are discussed. The TSNA are formed during tobacco processing and during pyrosynthesis, and may also be formed endogenously in the body. Their biochemical pathways and the evidence for their carcinogenicity in bioassays are presented together with a discussion of the research on utilizing TSNA as physiological markers of exposure to tobacco products.

3.1. Introduction

Nicotine was first isolated as a major constituent of tobacco in 1828;[1] however, it took 114 years to determine that this alkaloid was likely the major cause of tobacco habituation.[2] Fifty years ago, clinical data have incriminated cigarette smoking as a risk factor for cancer of the lung.[3,4] This observation was initially confirmed in 1950 by three epidemiological studies from the UK and the USA, respectively.[5–7] In the following 36 years, more than 200 epidemiological reports from at least 20 countries have formed the basis for the conclusions of the US Surgeon General, the Royal College of Physicians in London, and the International Agency for Research on Cancer, that cigarette smoking is causally associated with cancer of the lung, larynx, oral cavity, and oesophagus, and that cigarette smoking is correlated with cancer of the pancreas, kidney, and urinary bladder, and possibly with cancer of the cervix.[8–10] Cigar and pipe smoking are also causally associated with cancer of the lung although to a lesser degree than cigarette smoking.[8–10] More recently, passive smoking has been incriminated as a risk factor for cancer of the lung.[10] Furthermore, the oral use of snuff and the chewing of betel quid with tobacco have been associated with cancer of the oral cavity.[11,12]

The extensive epidemiological data on tobacco usage and cancer have led to the working hypothesis that all smoking products and smokeless tobaccos have one or more tobacco-specific factors in common which contribute greatly to the carcinogenic potential of these consumer products. On the basis of our present knowledge, we regard nicotine as a major tobacco-specific common denominator that is responsible for the carcinogenicity of chewing tobacco, snuff, and tobacco smoke.[13]

3.2. Nicotine and its metabolites as possible carcinogens

During the last four or five decades it has been thoroughly documented that nicotine metabolism occurs primarily in the liver by oxidation and, to some extent, by demethylation (Fig. 3.1).[14] α-Hydroxylation of nicotine leads to cotinine *via* the unstable 5'-hydroxynicotine. The latter exists in equilibrium with the nicotine-Δ-1' (5')iminium ion. This iminium ion is known to form adducts with glutathione. It has been hypothesized that this ion may also react with DNA and would thus be genotoxic in the mammalian cell.[15–18] However, at present there exists no evidence that the nicotine-Δ-1'(5')iminium ion is mutagenic or carcinogenic.

Administration of a major nicotine metabolite, cotinine, in the drinking water has led to malignant tumours in rats as reported by Truhaut *et al.*[19] These tumours were primarily lymphoreticular sarcomas in the large intestinal wall. Another group of investigators applied twice the dose of cotinine to BD-rats, but this assay did not lead to induction of tumours.[20]

In our own studies, we assayed *cis*- and *trans*-nicotine-*N'*-oxides and cotinine for tumour promoting activity in rats pretreated with the bladder carcinogen FANFT (N-[4-(5-nitro-2-furyl)-2-thiazolyl]formamide). Whereas cotinine, given in the drinking water for 18 months at the 0.1 per cent level was inactive as a tumour promoter, the data for the nicotine-*N'*-oxides were inconclusive. At the 0.02 per cent concentration, the nicotine-*N'*-oxides in the drinking water inhibited the formation of tumours in the bladder of FANFT-treated animals, however, both *N'*-oxides exhibited tumour promoting activity in the forestomach of the rats.[21] These findings require confirmation.

Low concentrations of nicotine (2.5–5 mg/ml) in solutions containing benzo(a)pyrene and TPA (tetradecanoylphorbol-13-acetate) increase the tumour formation on mouse skin.[22] This indicates cocarcinogenic activity for nicotine, but again, these results require confirmation.

3.3. Nicotine as a precursor to tobacco-specific nitrosamines

It is well established that secondary and tertiary amines can react *in vitro* and *in vivo* with nitrite, thus yielding nitrosamines.[23] More than 300 nitrosamines have been shown to be carcinogenic in one or more of 40 animal species.[24] Due to the presence of nornicotine (0.2–0.7 mg/g tobacco; 27–88 μg/smoke of one cigarette), anatabine (0.3.–1.0 mg/g; 4–20

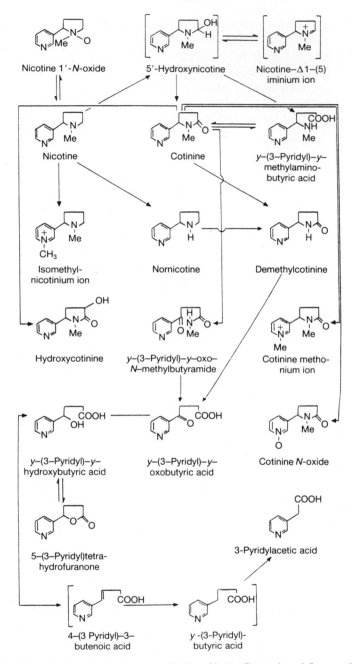

Fig. 3.1. Pathways of nicotine metabolism (from Gorrod and Jenner 1975).[14]

Table 3.1. Tobacco-specific nitrosamines in commercial US tobacco products[27,29]

Tobacco product*	NNN	NNK	NAT + NAB
A. Smokeless tobacco			
Chewing tobacco (ppb)	3500– 8200	100– 3000	500– 7000
Moist snuff (ppb)†	800–89 000	200– 8000	200–220 000
Dry snuff (ppb)†	9400–55 000	2000–14 400	20 000– 40 000
B. Mainstream smoke			
Cigarettes-NF (ng/cig.)	120–950	80–770	140–990
Cigarettes-F (ng/cig.)	50–310	30–150	60–370
Little cigar-F (ng/cigar)	5500	4200	1700
Cigar (ng/cigar)	3200	1900	1900
C. Sidestream smoke			
Cigarettes-NF (ng/cig.)	1700	410	270
Cigarettes-F (ng/cig.)	150	190	150

*NF, cigarette without filter tip; F, cigarette with filter tip.
†Iso-NNAL; moist snuff 66–2460 ppb; dry snuff 75–140 ppb.
Abbreviations: NNN, N'-nitrosonornicotine; NNK, 4-(methylnitrosamino)-1-butanone; NAT, N'-nitrosoanatabine; NAB, N-nitrosoanabasine; Iso-NNAL, 4-(methylnitrosamino)-4-(3-pyridyl)-1-butanol.

μg/cigarette) and anabasine (0.1–0.3 mg/g; 3–15 μg/cigarette) as secondary amines and with nicotine (1–21 mg/g; 800–3000 μg/cigarette) as a tertiary amine, tobacco, and its smoke have the potential to form tobacco-specific N-nitrosamines (TSNA) which may be carcinogenic. This concept had already been proposed by Druckrey and Preussmann in 1962.[25]

During the last decade, detailed studies have indeed shown that tobacco and tobacco smoke contain these TSNA (Fig. 3.2). These compounds are virtually non-existent in the green leaf at the time of harvesting (<1 ppb), but they are being formed during the curing and fermentation stages of tobacco processing and during ageing of the tobacco products. Nicotine gives rise to N'-nitrosonornicotine (NNN), 4-(methylnitrosamino)-1-(3-pyridyl)-1-butanone (NNK), 4-(methylnitrosamino)-1-(3-pyridyl)-1-butanol (NNAL), and to 4-(methylnitrosamino)-4-(3-pyridyl)-1-butanol (Iso-NNAL). Twenty to forty per cent of the NNN and NNK in cigarette smoke originates from tobacco by direct transfer into the smoke, the remainder is pyrosynthesized during the burning of tobacco.[26–29]

While nicotine gives rise to NNN, NNK, NNAL, and Iso-NNAL, the minor tobacco alkaloids nornicotine, anabasine, and anatabine are precursors to NNN, N'-nitrosoanabasine (NAB) and nitrosoanatabine (NAT). Table 3.1 presents analytical data for the concentrations of these TSNA in

28   *Dietrich Hoffmann*

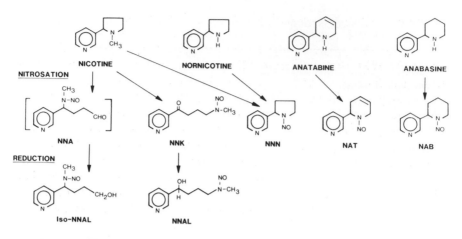

Fig. 3.2. Formation of tobacco-specific *N*-nitrosamines.

smokeless tobacco, mainstream smoke, and undiluted sidestream smoke. According to the US National Research Council, the daily exposure to TSNA of a smoker of 20 cigarettes exceeds by at least 10 times the exposure of a non-smoker to all environmental nitrosamines (Table 3.2).[30]

The concentrations of NNN and NNK formed during tobacco processing are profoundly influenced by the amounts of nicotine available and by the levels of nitrate in the leaf.[31] The curing and fermentation processes

Table 3.2. Estimated exposure of United States residents to nitrosamines* †

Source of exposure	Nitrosamines	Primary exposure route	Daily intake (μg/person)
Beer	NDMA	Ingestion	0.34
Cosmetics	NDELA	Dermal absorption	0.41
Cooked meat, cooked bacon	NPYR	Ingestion	0.17
Scotch whiskey	NDMA	Ingestion	0.03
Cigarette Smoking	VNA	Inhalation	0.3
	NDELA	Inhalation	0.5
	NNN	Inhalation	6.1 ⎫
	NNK	Inhalation	2.9 ⎬ 16.2‡
	NAT + NAB	Inhalation	7.2 ⎭

*National Research Council, 1981.[30]

†Abbreviations: NDMA, nitrosodimethylamine; NDELA, nitrosodiethanolamine; NPYR, nitrosopyrrolidine; VNA, volatile nitrosamines; abbreviations for NNN, NNK, NAT, NAB see Table 3.1.

‡Sum of tobacco-specific nitrosamines.

applied in the preparation of the tobacco product also influence the formation of these tobacco-specific *N*-nitrosamines and, as stated before, ageing of the processed tobacco yields additional amounts of nitrosamines.[26,32]

The effects of nitrate levels in tobacco on the degree of NO_x formation during smoking and the resulting smoke yields of *N*-nitrosamines are shown in Fig. 3.3.[26] Over a period of 25 years the nitrate content of the US blended cigarette has risen from about 0.5 to 1.2–1.5 per cent.[8] This increase is primarily due to the use of Burley tobacco varieties for the purpose of creating cigarettes with low smoke yields, and to the use of about 20 per cent of ribs and stems in tobacco blends. In a study with several experimental cigarettes made with 70 per cent of the same blend of laminae and 30 per cent of various types of ribs, we demonstrated that the nitrate content of the ribs greatly affects the yields of TSNA in mainstream smoke ($r^2 = 0.922$) and in sidestream smoke ($r^2 = 0.767$).[33]

In order to estimate the contribution of nornicotine and nicotine in the tobacco to the *N'*-nitrosonornicotine yield in the smoke, we enriched the nornicotine and nicotine content of a blended US cigarette. Water solutions of nornicotine tartrate or nicotine tartrate, respectively, were injected into the tobacco column with a microsyringe and the cigarettes were then smoked under standard conditions.

As shown in Tables 3.3 and 3.4, an eight-fold increase in nornicotine from 0.73 to 5.92 mg in the tobacco smoked results only in a 50 per cent

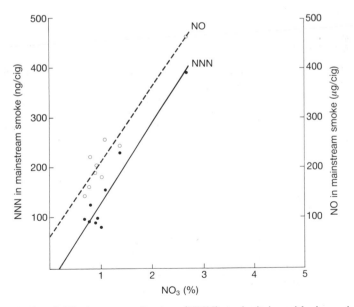

Fig. 3.3. Yields of *N'*-nitrosonornicotine (NNN) and nitric oxide in mainstream smoke as a function of percentage of nitrate in tobacco.[26]

Table 3.3. Nitrosonornicotine (NNN) in the smoke of cigarettes enriched with nornicotine*

Nornicotine added (mg/cig)	Nornicotine in total tobacco smoked (mg/cig)	Mainstream smoke (mg/cig)	Nornicotine transfer rate (%)	Burning rate (mg/min)	TPM† (dry) (mg/cig)	NNN (ng/cig)
0	0.73	0.03	4.1	59.1	28.0	137
1.87	2.60	0.07	2.7	63.0	27.6	129
4.11	4.84	0.12	2.5	65.8	28.1	168
5.15	5.92	0.18	3.0	66.3	27.7	214

*Added to the tobacco as nornicotine tartrate.
†Total particulate matter.
Hoffmann et al. (1974).[52]

increase of NNN, whereas doubling of nicotine from 14.8 to 29.7 mg results in three times the amount of NNN in the smoke (Fig. 3.4).[52] Since the ratio of nicotine to nornicotine in most tobaccos is of the order of at least 15:1 or even 20:1, this result strongly incriminates nicotine as the major precursor for N'-nitrosonornicotine (and NNK) in cigarette smoke. Thus, a significant reduction in the concentration of nicotine in the tobacco will lead to a significant reduction of the tobacco-specific N-nitrosamines in the smoke.

As in the case of plain cigarettes, the nicotine and nitrate contents of the tobacco blend are also determining factors for the smoke yield of filter cigarettes. While the cellulose acetate filter tips are highly selective in removing volatile nitrosamines from the smokestream,[34] the removal of

Table 3.4. Nitrosonornicotine (NNN) in the smoke of cigarettes enriched with nicotine*

Nicotine added (mg/cig)	Nicotine in total tobacco smoked (mg/cig)	Mainstream smoke (mg/cig)	Nicotine transfer rate (%)	Burning rate (mg/min)	TPM† (dry) (mg/cig)	NNN (ng/cig)
0	14.8	1.88	12.8	59.1	27.4	137
5.5	20.3	2.30	11.3	59.2	27.9	219
10.1	24.9	2.70	10.8	59.4	28.1	374
14.3	29.1	3.20	10.9	61.2	28.4	393

*Added to the tobacco as nicotine tartrate. No significant increase in nornicotine was found in the mainstream smoke of the nicotine-enriched tobacco.
†Total particulate matter.
Hoffmann et al., (1974).[52]

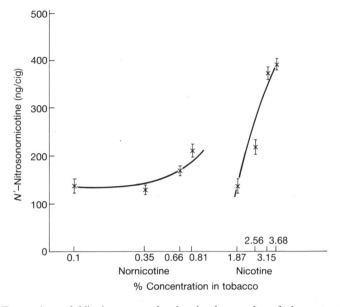

Fig. 3.4. Formation of *N'*-nitrosonornicotine in the smoke of cigarettes enriched with nornicotine or nicotine.[52]

TSNA by filter tips parallels that of the 'tar' in general and is thus not selective.[26] Furthermore, our studies do not indicate that smoke dilution by perforated filter tips leads to a higher reduction for TSNA than for 'tar'.[26] This is in line with the observation that nicotine is not selectively reduced by air dilution.[35] We explored the possibility of inhibiting the formation of TSNA during cigarette smoking by adding 0.1 per cent ascorbic acid to tobacco ribs; however, this approach was not effective.[33]

An important question is whether NNN and NNK can be formed by endogenous nitrosation of nicotine following cigarette smoke inhalation. The major problem we face here lies in the difficulty in differentiating between NNN and NNK which is inhaled with the smoke, and the NNN and NNK conceivably formed endogenously upon smoke inhalation. The short half-life of these nitrosamines *in vivo* adds to the difficulty of their assessment.[43] We know that a cigarette smoker has a significantly greater potential for endogenous nitrosamine formation than a non-smoker from comparisons of the urinary excretion of nitrosoproline (NPRO) by cigarette smokers and non-smokers.[36-40] Endogenous nitrosation of proline had earlier been demonstrated with high-nitrite diets by Ohshima and Bartsch.[53] Since NPRO is not metabolized and is excreted exclusively in the urine, its concentration in urine is a reliable measure for the degree of endogenous nitrosation of amines.

Table 3.5 summarizes our studies on the increased endogenous formation of NPRO in cigarette smokers.[36] Male smokers and male non-smokers were placed on a controlled diet for 12 days. On day 3, we collected the urine and found significantly higher excretion of NPRO by smokers than by non-smokers. When proline was added to the diet, the urinary NPRO excretion increased further in smokers, but remained the same in non-smokers. When proline and the nitrosation inhibitor ascorbic acid were added to the diet, the urinary excretion of NPRO in smokers was at the same level as that of non-smokers. This assay demonstrates that cigarette smoking increases the endogenous formation of nitrosamines in man. Subsequent to this assay, four additional reports have shown that the excretion of NPRO as well as that of nitrosothioproline is significantly increased in cigarette smokers compared to non-smokers.[37–40] Although we may assume that nicotine gives rise to the endogenous formation of NNN and NNK in tobacco users, at present we are lacking direct evidence for this occurrence.

Table 3.5. Endogenous formation of *N*-nitrosoproline in man (μg/24-hour urine void)

Protocol	Non-smokers	N	Smokers	N
Control diet	3.6 ± 2.1	13	5.9	13
Diet + proline	3.6	14	11.8	14
Diet + proline + vitamin C	4.7	13	4.6	13
Diet + vitamin C	4.0	9	6.0	8

Hoffmann and Brunnemann (1983).[36]

3.4. Carcinogenicity of tobacco-specific *N*-nitrosamines

In bioassays of the six identified tobacco-specific nitrosamines we found NNN and NNK to be powerful carcinogens in mice, rats, and hamsters.[27,41] A single dose of 1 mg of NNK induces a significant incidence of tumours in the respiratory tract of hamsters[44] and, on a molar basis, NNK is a more powerful carcinogen in rats than *N*-nitrosodimethylamine.[45] Recently, Boutet et al.[49] have shown that intratracheal instillation into hamsters of a single dose of 2 mg of NNK induces preneoplastic changes similar to those observed in the lung tissue of smokers within the first 3 months.[50,51]

Swabbing of the oral cavity of rats with a solution containing a mixture at low doses of NNN and NNK induced tumours of the mouth in 8 of 30 animals, and tumours of the lung in 5 of 30 rats, indicating that these nitrosamines are active as contact carcinogens as well as organ-specific carcinogens.[41] Activity as a contact carcinogen was also observed for NNK though not for NNN, when these compounds were individually tested as tumour initiators on mouse skin.[42] NAB and NNAL are moderately active carcinogens; Iso-NNAL, which was only recently identified, is now being tested for tumour genicity. NAT was inactive in rats at dose levels up to 9 mmol/kg (Table 3.6).

3.5. Biochemistry of tobacco-specific *N*-nitrosamines

The high concentration of the nitrosamines derived from the *Nicotiana* alkaloids in tobacco products, their tobacco-specificity and their high carcinogenic activity required studies on their biochemical effects in mammalian systems.[27]

Figure 3.5 summarizes the biochemical pathways of NNK activation *in vitro* in animal and human tissues, and *in vivo* in mice, rats, and hamsters. α-Hydroxylation of this unsymmetrical dialkylnitrosamine leads to two highly unstable α-hydroxynitrosamines. These decompose with formation of 4-(3-pyridyl)-4-oxobutanal and formaldehyde, as well as two unstable diazohydroxides, namely methyl diazohydroxide and 4-(3-pyridyl)-4-oxobutyldiazohydroxide. Methyldiazohydroxide is a known methylating agent. When rats were treated with NNK, 7-methylguanine, O^6-methylguanine, and O^4-methylthymidine were formed in the DNA of lung, liver, and nasal mucosa.[45–47] All of these are organs in which NNK induces tumours. Studies with cultured human buccal mucosa, trachea, oesophagus, bronchus, peripheral lung, and bladder have shown that these

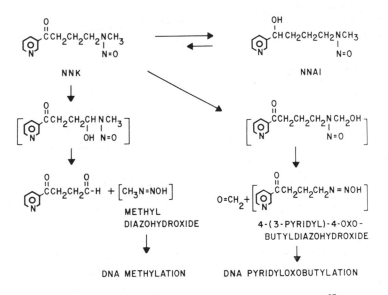

Fig. 3.5. Metabolic pathways of NNK activation.[27]

Dietrich Hoffmann

Table 3.6. Carcinogenicity of tobacco-specific nitrosamines (TSNA)*†

Nitrosamine	Species and strains	Route of application	Principal target-organs	Dose
NNK	A/J	i.p.	lung	0.12 mmol/mouse
	F344 rat	s.c.	nasal cavity, liver, lung	0.1–2.8 mmol/rat
	Syrian golden hamster	s.c.	trachea, lung, nasal cavity	0.005–0.9 mmol/hamster
NNN	A/J mouse	i.p.	lung	0.12 mmol/mouse
	F344 rat	s.c.	nasal cavity, oesophagus	0.2–3.4 mmol/rat
	F344 rat	p.o.	nasal cavity, oesophagus	1.0–3.6 mmol/rat
	Sprague–Dawley rat	p.o.	nasal cavity	8.8 mmol/rat
	Syrian golden hamster	s.c.	trachea, nasal cavity	0.9–2.1 mmol/hamster
NAT	F344 rat	s.c.	none	0.2–2.8 mmol/rat
NAB	F344 rat	p.o.	oesophagus	3–12 mmol/rat
	Syrian golden hamster	s.c.	none	2 mmol/mouse
NNA	A/J mouse	i.p.	none	0.12 mmol/mouse
NNAL	A/J mouse	i.p.	lung	0.12 mmol/mouse
Iso-NNAL	F344 rat	s.c.	? under test	0.12 mmol/mouse

*Hoffmann and Hecht (1985).[27]
†Abbreviations for TSNA: NNK, 4-(methylnitrosamino)-1-(3-pyridyl)-1-butanone; NNN, *N'*-nitrosornicotine; NAT, *N'*-nitrosoanatabine; NAB, *N'*-nitrosoanabasine; NNA, 4-(methylnitrosamino)-4-(3-pyridyl)butanal; NNAL, 4-(methylnitrosamino)-1-(3-pyridyl)-1-butanol; Red NNAL 4-(methylnitrosamino)-4-(3-pyridyl)butanol.

tissues too can metabolize NNK by α-hydroxylation.[46] Therefore, it is reasonable to expect that tobacco smoking will lead to the formation of methylated bases in the DNA of human tissues. O^6-Methylguanine and O^4-methylthymidine have been shown to cause miscoding of DNA and are important in regard to tumour initiation. Figure 3.6 presents a scheme showing the link of the major and habituating tobaco alkaloid, nicotine, with the promutagenic DNA adducts O^6-methylguanine and O^4-methylthymidine.[27]

Fig. 3.6. Scheme linking nicotine to formation of the promutagenic DNA adducts O^6-methylguanine and O^4-methylthymidine.

As we discussed, the α-hydroxylation of NNK at the N-methyl group leads to the alkylating agent 4-(3-pyridyl)-4-oxobutyldiazohydroxide. The same intermediate is formed upon α-hydroxylation of NNN.[27] We are currently exploring the DNA adduct(s) of this alkylating metabolite *in vitro* and *in vivo*. The isolation of such DNA adduct(s) will be of major significance because of the specificity of their origin from tobacco. After all, they derive strictly from nicotine *via* tobacco-specific *N*-nitrosamines.

3.6. Tobacco-specific *N*-nitrosamines as biological markers

Due to the specificity of NNN and NNK as tobacco carcinogens and because of our knowledge about their metabolic fate, we favour their use as indicators for the uptake of tobacco smoke and environmental tobacco smoke. As discussed before, α-hydroxylation of NNK at the N-methyl group leads to the alkylating agent 4-(3-pyridyl)-4-oxobutyldiazohydroxide. The same intermediate is also formed upon α-hydroxylation of NNN.[27] Recently, Stephen Hecht from our group has demonstrated that [5-³H]NNN and [5-³H]NNK bind to globin. Hydrolysis of the globin adduct enables isolation of 4-hydroxy-1-(3-pyridyl)-1-butanone (Fig. 3.7).[48] Following reduction with tritium to the corresponding secondary alcohol, HPLC separation allows isolation of femtomoles of this metabolite of NNN and NNK. This method is now being explored for dosimetric studies on cigarette smokers which require only 10–20 ml of blood. We are hopeful that this technique will allow determination of the uptake of the tobacco-specific *N*-nitrosamines by cigarette smokers, respectively, of the amount of NNN and NNK, present due to their uptake and endogenous formation.

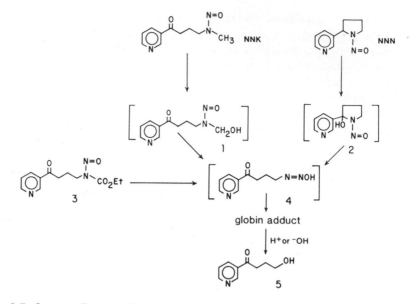

Fig. 3.7. Intermediates and products involved in the binding to NNK, NNN, and 4-(carbethoxynitrosamino)-1-(3-pyridyl)-1-butanone.[48]

3.7. Summary

Whereas it has long been known that nicotine is a major habituating agent in smokeless tobacco and in tobacco smoke, studies during the last decade have clearly demonstrated that nicotine gives rise to two powerful carcinogenic *N*-nitrosamines, namely NNN and NNK. These agents induce benign and malignant tumours in the oesophagus and the respiratory tract, including the lungs of mice, rats, and hamsters.[49] Other newly identified nitrosamines also deriving from nicotine are as yet being assayed for carcinogenic activity.

Biochemical studies have clearly demonstrated that metabolic activation of NNN and NNK leads to the formation of mutagenic DNA adducts. Such adducts are potentially useful as indicators of the uptake of tobacco smoke and environmental tobacco smoke. Such adducts are also formed *in vitro* in tissues deriving from human target organs that are known to be at increased risk for cancer in cigarette smokers. On the basis of these facts and because the concentration of these nicotine-derived nitrosamines represents the far highest levels of carcinogenic nitrosamines ever reported in an inhalant, we strongly support efforts leading to further reduction of nicotine in all tobacco products. We contend that biochemical studies are the only ones capable of establishing a relationship between cancer risk and a specific group of compounds when these compounds are major carcinogenic constituents of a complex matrix such as tobacco smoke with more than 4000 compounds.

Acknowledgements

I greatly appreciate the inspiration and support given by Ernst L. Wynder, the founder of the American Health Foundation, and the extensive contributions of my colleagues John D. Adams, Klaus D. Brunnemann, Fung-Lung Chung, Stephen S. Hecht, Edmond J. LaVoie, and Abraham S. Rivenson. I thank Bertha Stadler and Ilse Hoffmann for their editorial assistance.

Our studies on tobacco are supported by Grant CA-29580 from the United States National Cancer Institute.

References

1. Posselt, W. and Reimann, L. (1828). Chemische Untersuchungen des Tabaks und Darstellung eines eigentümlichen wirksamen Prinzips dieser Pflanze. *Geigers Magazin für Pharmacologie*, **24**, 138–41.
2. Johnston, L. M. (1942). Tobacco smoking and nicotine. *Lancet*, **ii**, 742.
3. Müller, F. H. (1939). Tabakmissbrauch und Lungencarcinom. *Zeitschrift für Krebsforschung*, **49**, 57–84.
4. Schairer, E. and Schoeninger, E. (1943). Lungenkrebs und Tabakverbrauch. *Zeitschrift für Krebsforschung*, **54**, 261–9.
5. Doll, R. and Hill, A. B. (1950). Smoking and carcinoma of the lung. Preliminary report. *British Medical Journal*, **2**, 739–48.
6. Wynder, E. L. and Graham, E. A. (1950). Tobacco smoking as a possible aetiological factor in bronchiogenic carcinoma. A study of six hundred and eighty-four proven cases. *Journal of the American Medical Association*, **143**, 329–36.
7. Levin, M. L., Goldstein, H., and Gerhardt, P. R. (1950). Cancer and tobacco smoking. A preliminary report. *Journal of the American Medical Association*, **143**, 336–8.
8. US Surgeon General (1982). *The health consequences of smoking – cancer.* Report of the Surgeon General's Advisory Committee, US Public Health Service DHHS Publication Number (PHS) 82–50179, US Government Printing Office, Washington, DC.
9. Royal College of Physicians, London (1983). *Health or smoking?* Follow-up Report of the Royal College of Physicians. Pitman, London.
10. International Agency for Research on Cancer (1986). *Tobacco smoking*, Monograph No. 38. IARC, Lyon, France.
11. International Agency for Research on Cancer (1985). *Tobacco habits other than smoking. Betel-quid and areca-nut chewing: and some related nitrosamines*, Monograph No. 37. IARC, Lyon, France.

12. US Surgeon General (1986). *The health consequences of using smokeless tobacco*, A report of the Advisory Committee to the Surgeon General. NIH Publication 82–2874.

13. Hoffmann, D., LaVoie, E. J., and Hecht, S. S. (1985). Nicotine: a precursor for carcinogens. *Cancer Letters*, **28**, 67–75.

14. Gorrod, J. W. and Jenner, P. (1975). The metabolism of tobacco alkaloids. *Essays in Toxicology*, **6**, 35–78.

15. Sanders, E. B., DeBardeleben, J. F., and Osdene, T. S. (1975). Nicotine chemistry. 5′-Cyanonicotine. *Journal of Organic Chemistry*, **40**, 2848–9.

16. Nguyen, T. R., Gruenke, L. D., and Castagnoli, J., Jr (1979). Metabolic oxidation of nicotine to chemically reactive intermediates. *Journal of Medicinal Chemistry*, **22**, 259–63.

17. Hibbard, A. R. and Gorrod, J. W. (1982). Nicotine-$\Delta 1'(5')$iminium ion: a reactive intermediate in nicotine metabolism. *Advances in Experimental Medicine and Biology*, **136B**, 1121–31.

18. Overton, M., Hickman, J. A., and Threadgill, M. D. (1985). The generation of potentially toxic, reactive iminium ions from the oxidative metabolism of xenobiotic N-alkyl compounds. *Biochemical Pharmacology*, **34**, 2055–61.

19. Truhaut, R., DeClercq, M., and Loisillier, F. (1964). Acute and chronic toxicity of cotinine and its carcinogenic activity in the rat. *Pathologie Biologie*, **12**, 39–42.

20. Schmähl, D. and Osswald, H. (1968). Fehlen einer carcinogenen Wirkung von Cotinin bei Ratten. *Zeitschrift für Krebsforschung*, **71**, 198.

21. LaVoie, E. J., Shigematsu, A., Rivenson, A., Mu, B., and Hoffmann, D. (1985). Evaluation of the effects of cotinine and nicotine-N-oxides on the development of tumors in rats initiated with N-(4-(5-nitro-2-furyl)-2-thiazolyl) formamide. *Journal of the National Cancer Institute*, **75**, 1075–81.

22. Bock, F. G. (1980). Cocarcinogenic properties of nicotine. In *Banbury Report, 3 – A safe cigarette* (ed. G. Gori and F. Bock), pp. 129–39. Cold Spring Harbor Laboratory, New York.

23. Mirvish, S. S. (1975). Formation of N-nitroso compounds: chemistry, kinetics, and *in vivo* occurrence. *Toxicology and Applied Pharmacology*, **31**, 325–51.

24. Preussmann, R. and Stewart, B. W. (1984). N-Nitroso carcinogens. *American Chemical Society Monographs*, **182**, 643–828.

25. Druckrey, H. and Preussmann, R. (1962). Zur Entstehung carcinogener Nitrosamine am Beispiel des Tabakrauches. *Naturwissenschaften*, **49**, 498–9.

26. Hoffmann, D., Brunnemann, K. D., Adams, J. D., and Hecht, S. S. (1984). Formation and analysis of N-nitrosamines in tobacco products and their endogenous formation in tobacco consumers. *IARC Scientific Publications*, **57**, 743–62.

27. Hoffmann, D. and Hecht, S. S. (1985) Perspectives in cancer research. Nicotine-derived N-nitrosamines and tobacco-related cancer: Current status and future directions. *Cancer Research*, **45**, 934–44.

28. Brunnemann, K. D., Genoble, L., and Hoffmann, D. (1987). Identification and analysis of a new tobacco-specific N-nitrosamine, 4-(methylnitrosamino)-4-(3-pyridyl)-1-butanol. *Carcinogenesis*, **8**, 465–9.

29. Hoffmann, D., Adams, J. D., Lisk, D., Fisenne, I., and Brunnemann, K. D. (1987). Toxic and carcinogenic agents in dry and moist snuff. *Journal of the National Cancer Institute*, **79**, 1281–6.

30. National Research Council (1981). *N*-Nitroso Compounds: Environmental Distribution and Exposure of Humans. In *The health effects of nitrate, nitrite, and N-nitroso compounds*, Part 1, pp. 7–36. National Academy Press, Washington, DC.

31. Brunnemann, K. D., Scott, J. C., and Hoffmann, D. (1983). N-nitrosoproline, an indicator for N-nitrosation of amines in processed tobacco. *Journal of Agricultural and Food Chemistry*, **31**, 905–9.

32. Hoffmann, D. and Adams, J. D. (1981). Carcinogenic tobacco-specific N-nitrosamines in snuff and in the saliva of snuff dippers. *Cancer Research*, **41**, 4305–8.

33. Brunnemann, K. D., Masaryk, J., and Hoffmann, D. (1983). The role of tobacco stems in the formation of N-nitrosamines in tobacco and cigarette mainstream and sidestream smoke. *Journal of Agricultural and Food Chemistry*, **31**, 1221–4.

34. Brunnemann, K. D., Yu, L., and Hoffmann, D. (1977). Assessment of carcinogenic volatile N-nitrosamines in mainstream and sidestream smoke from cigarettes. *Cancer Research*, **37**, 3218–22.

35. Norman, V. (1974). The effect of perforated tipping paper on the yields of various smoke components. *Beiträge zur Tabakforschung*, **7**, 282–7.

36. Hoffmann, D. and Brunnemann, K. D. (1983). Endogenous formation of N-nitrosoproline in cigarette smokers. *Cancer Research*, **43**, 5570–4.

37. Ladd, K. F., Newmark, H. L., and Archer, M. C. (1984). N-nitrosation of proline in smokers and nonsmokers. *Journal of the National Cancer Institute*, **73**, 83–7.

38. Tsuda, M., Nutsuma, J., Sato, S., Hirayama, F., Kakizoe, T., and Sugimura, T. (1986). Increase in the levels of N-nitrosoproline, N-nitrosothioproline, and N-nitroso-2-methylthioproline in human urine by cigarette smoking. *Cancer Letters*, **30**, 117–24.

39. Lu, S. H., *et al.* (1986). Urinary excretion of N-nitrosamino acids and nitrate by inhabitants of high- and low-risk areas for oesophageal cancer in Northern China: endogenous formation of nitrosoproline and its inhibition by vitamin C. *Cancer Research*, **46**, 1485–91.

40. Scherer, G. and Adlkofer, F. (1986). Endogenous formation of N-nitrosoproline in smokers and nonsmokers. In *Banbury Report 3 – A safe cigarette* (ed. G. Gori and F. Bock), pp. 137–48. Cold Spring Harbor Laboratory, New York.

41. Hecht, S. S., Rivenson, A., Braley, J., DiBello, J., Adams, J. D., and Hoffmann, D. (1986). Induction of oral cavity tumors in F344 rats by tobacco-specific nitrosamines and snuff. *Cancer Research*, **46**, 4162–6.

42. LaVoie, E. J., Prokopczyk, G., Rigotty, J., Czech, A., Rivenson, A., and Adams, J. D. (1987). Tumorigenic activity of the tobacco-specific nitrosamines 4-(methylnitrosamino)-1-(3-pyridyl)-1-butanone (NNK), 4-(methyl-nitrosamino)-4-(3-pyridyl)-1-butanol (iso-NNAL) and N'-nitrosonornicotine (NNN) on topical application to Sencar mice. *Cancer Letters*, **37**, 277–83.

43. Adams, J. D., LaVoie, E. J., O'Mara-Adams, K. J., and Hoffmann, D. (1985). Pharmacokinetics of N'-nitrosonornicotine and 4-(methyl-nitrosamino)-1-(3-pyridyl)-1-butanone in laboratory animals. *Cancer Letters*, **28**, 195–201.

44. Hecht, S. S., Adams, J. D., Numoto, S., and Hoffmann, D. (1983). Induction of respiratory tract tumors in Syrian golden hamsters by a single dose of

4-(methylnitrosamino)-1-(3-pyridyl)-1-butanone (NNK) and the effect of smoke inhalation. *Carcinogenesis*, **4**, 1287–90.

45. Hecht, S. S., Trushin, N., Castonguay, A., and Rivenson, A. (1986). Comparative carcinogenicity and DNA methylation in F344 rats by 4-(methylnitrosamino)-1-(3-pyridyl)-1-butanone and 4-nitrosodimethylamine. *Cancer Research*, **46**, 498–502.

46. Castonguay, A., Stoner, G. D., Schut, H. A. J., and Hecht, S. S. (1983). Metabolism of tobacco-specific N-nitrosamines by cultured human tissues. *Proceedings of the National Academy of Sciences of the United States of America*, **80**, 6694–7.

47. Belinsky, S. A., White, C. M., Boucheron, J. A., Richardson, F. C., Swenberg, J. A., and Anderson, M. (1986). Accumulation and persistence of DNA adducts in respiratory tissue of rats following multiple administration of the tobacco-specific carcinogen 4-(N-methyl-N-nitrosamino)-1-(3-pyridyl)-1-butanone. *Cancer Research*, **46**, 1280–4.

48. Hecht, S. S., *et al.* (1987). Investigations on the molecular dosimetry of tobacco-specific N-nitrosamines, *IARC Scientific Publications*, **84**, 423–9.

49. Boutet, M., Montiminy, L., and Castonguay, A. (1987). Cellular changes induced by the tobacco-specific carcinogen NNK in the respiratory tract of Syrian golden hamsters. *IARC Scientific Publications*, **84**, 438–42.

50. Auerbach, O. *et al.* (1957). Changes in the bronchial epithelium in relation to smoking and cancer of the lung. *New England Journal of Medicine*, **256**, 97–104.

51. Spencer, H. (Editor) (1985). Pre-carcinomatous conditions in the lung. In *Pathology of the lung*, 4th edn, (ed. H. Spencer), pp. 851–932. Pergamon Press, New York.

52. Hoffmann, D., Rathkamp, G., and Liu, Y. Y. (1974). On the isolation and identification of volatile and non-volatile N-nitrosamines and hydrazines in cigarette smoke. *IARC Scientific Publications*, **9**, 159–65.

53. Ohshima, H. and Bartsch, H. (1981). Quantitative estimation of endogenous nitrosation in humans by monitoring N-nitrosoproline excreted in the urine. *Cancer Research*, **41**, 3658–62.

54. Gorrod, J. W. and Jenner, P. (1975). The metabolism of tobacco alkaloids. *Essays in Toxicology*, **6**, 35–78.

4

The toxicity of nicotine: cancer

FRANCIS J. C. ROE

Abstract

The general toxicology of nicotine is briefly reviewed. It is not tera-togenic, genotoxic, or carcinogenic per se. The nitrosation of nicotine derivatives, either during the curing of tobacco or by pyrolysis during smoking, gives rise to nitrosamines which have been shown to be carcinogenic in animals. The contribution, if any, of such nitrosamines in tobacco to cancer risk in smokers is uncertain for several reasons. In particular, the strikingly lower lung cancer risk in pipe smokers compared with cigarette smokers, despite the presence of the same nitrosamines in pipe smoke, argues against their importance. Several constituents of smoke have been reported to be carcinogenic for laboratory animals. There is no firm scientific basis for believing that nitrosamines derived from nicotine are of paramount importance in relation to carcinogenic risks associated with smoking. In so far as the nicotine content of smoke contributes significantly to the enjoyment that most smokers derive from smoking, it would seem wiser, on the basis of presently available knowledge, to increase the ratio of nicotine to other smoke constituents rather than to reduce it. In Britain the yields of nitrosamines in cigarette smoke are very low because of the selective use of low nitrate tobaccos, filtration, and ventilation in cigarette design.

4.1. Introduction

This paper reviews and discusses the toxicology of nicotine with particular reference to carcinogenic risk. It does not discuss the toxicology or carcino-genicity of tobacco or tobacco smoke. Like many other substances that have been around for a long time, there is not a plethora of data derived from toxicity tests of modern design. Nevertheless, there are arguably enough data to provide the basis for a common-sense view. It is perhaps relevant to point out that in the recent IARC Monograph on Tobacco Smoking,[1] the word 'nicotine' appears only once in the four-page long index, and this is in relation to its concentration in smoke and not in relation to the assessment of its toxicity. The fact is that this 394-page

41

volume is not concerned with nicotine *per se*. Furthermore, nicotine is not listed among the 700+ substances for which the IARC has prepared monographs.

The question of whether or not nicotine *per se* is a procarcinogen (i.e. a compound which is itself not carcinogenic, but which can be converted into a carcinogen in the body by metabolic processes) has more than just academic interest. Basically, there are three possibilities. The first is that tobacco smoke consists of a non-carcinogenic and non-procarcinogenic substance, nicotine, in a carcinogenically active vehicle. The second is that nicotine is an integral part of the carcinogenic activity of smoke. And the third is that nicotine is an important procarcinogen in smoke. If the first proposition is true, then the development of safe products will depend on selectively decreasing exposure to the vehicle. If the second is true then there can be little hope of developing safer products and no case for changing the vehicle: nicotine ratio. However, if the third is true, then we should be concentrating on reducing exposure to nicotine. The way in which most smokers smoke a cigarette is largely determined by their desire to obtain a particular dose of nicotine. Consequently, if a smoker is given a cigarette with a lower nicotine delivery than his usual cigarette, he will tend to smoke a greater number until he gets close to his usual dose of nicotine. In the process, he may get also his usual dose of tar or an even higher dose of tar than usual. The simple question is, therefore, which is likely to be safer, a low nicotine/medium tar cigarette, a low nicotine/low tar cigarette, or a medium nicotine/low tar cigarette?

4.2. Assessment of the toxicity of nicotine

The usual first step in assessing the toxicity of a substance is to look at its chemical structure, the way it is metabolized, and the data from acute toxicity studies. Structurally, nicotine is a pharmacologically active alkaloid capable of binding to specific receptor sites, but lacking in alkylating activity. It is convertible to nornicotine which, like many pharmacologically active substances, is a secondary amine susceptible to nitrosation. *In vivo* nicotine is, for the main part, rapidly metabolized to cotinine which is almost completely lacking in pharmacological activity.[2]

Of the many minor metabolites the only ones of possible theoretical interest in relation to mutagenic or carcinogenic risk are the *N*-oxides of nicotine and cotinine. However, for neither of these substances are there any substantial toxicological data.

There exist acute toxicity data for nicotine itself, both from animal studies (Table 4.1) and for man (Table 4.2). These indicate that nicotine is a relatively potent toxin, but are otherwise unremarkable.

Table 4.1. Acute toxicity of nicotine: animals

Species	Route	LD_{50} (mg/kg)	Reference
Mouse	oral	3.3–35	3–5
	i.v.	0.55–7.1	3–6
Rat	oral	50–188	3,4
	i.v.	2.8	4
Dog	oral	9.2	5
	i.v.	5	3,4

As is the case for many other substances that have been around for a long time there are no data available from chronic toxicity animal studies of modern design. Such data as there are, with one exception, are consistent with nicotine being without carcinogenic activity. The one exception is the report of Staemmler[7] of the occurrence of adrenal medullary tumours in rats as a consequence of their exposure to nicotine. We now know that this form of neoplasia is readily induced in rats by simple overfeeding and by the dietary administration of lactose or polyols which greatly enhance the absorption of calcium from the rat gut.[8] It is probable that the tumours which Staemmler reported were either a consequence of the catecholamine releasing activity of nicotine or a non-specific consequence of the experimental circumstances in which the observations were made. Further studies of modern design under carefully controlled conditions would be needed to distinguish between these two possibilities.

There is no clear-cut published evidence to indicate that the concentration of nicotine *per se* in inhaled smoke influences the risk of any form of cancer in animals or man. A problem is that it is difficult to change the yield of nicotine in smoke without changing the yields of many other smoke components. For this reason there are no hard data. However, the fact that pipe smoking, despite substantial uptake of nicotine, carries only a very low level of risk for lung cancer provides substantial support for the view that

Table 4.2. Acute toxicity of nicotine: man

General—headache, sweating, palpitation, fatigue
Gastro-intestinal—nausea, vomiting, abdominal pain, diarrhoea
Information from suicides who ingested nicotine-containing insecticides
Small oral doses better tolerated by smokers than by non-smokers
0.5–1 mg/kg as a single dose can be lethal
However, cumulative oral doses of up to 4 mg/kg can be taken in the form of
    tobacco during 24 hours without obvious toxicity

neither nicotine itself, nor its metabolites, nor nitrosamines derived from tobacco alkaloids are significant lung carcinogens. It is commonly held that pipe smokers are only at low risk of lung cancer because they do not inhale or because they only inhale a little. Personal observation of pipe smokers, however, suggests that many of them do, in fact, take smoke into their lower respiratory tracts. If any event, pipe smokers can certainly build up substantial levels of nicotine in their circulating blood. This is partly because nicotine is more easily absorbed from neutral or slightly alkaline pipe smoke than from the more acid smoke of flue-cured cigarettes. However, it maybe also partly due to absorption from the lower respiratory tract. For the purposes of the present argument it does not really matter how the nicotine gets into pipe smokers − if they are substantially dosed with nicotine and do not get cancer, then this argues rather strongly against the concept that nicotine contributes seriously to the carcinogenicity of smoke.

The US Surgeon General in his 1981 report[9] concluded that nicotine is not a complete carcinogen nor a tumour promoter. However, he went on to opine, on the basis of mouse skin painting studies, that nicotine is an important co-carcinogen. According to Cohen and Roe,[2] however, the evidence is equivocal. Experiments designed to see whether nicotine influences skin carcinogenesis in mice have, in fact, given mixed results − sometimes enhancement and sometimes inhibition.

The role of nicotine in the causation of particular diseases in man is also problematical. It has been suggested that nicotine lowers the threshold for ventricular fibrillation, that it strains the heart by increasing blood pressure and heart rate (i.e. through catecholamine release) and that it favours thrombosis by increasing platelet stickiness and favouring platelet aggregation.[2] However, the picture is not clear-cut in relation to any of these effects. Furthermore, although nicotine has the effect of increasing serum fatty acids, it has no effect on serum triglycerides or cholesterol. For this reason and in the light of other evidence it is not seriously suspected of implication in the causation of atherosclerosis.[10] Finally, there is no reason to suspect that nicotine *per se* is implicated in any association between smoking and peptic ulceration. On the other hand, it is plausible that nicotine, through some consequence of its neuropharmacological activity, is responsible for the relative dearth of cases of Parkinson's disease in smokers. Whether the reduced risk of ulcerative colitis in smokers as compared with non-smokers is attributable to nicotine is uncertain. Anecdotal evidence suggests nicotine is responsible.[11] However, a later study did not confirm this.[12]

As far as teratogenicity and embryotoxicity are concerned (Table 4.3), smoking is associated with decreased birth weight, but not with congenital deformities in man. In animals, no relevant data from studies involving

Table 4.3. Evaluation of nicotine for teratogenicity and embryotoxicity

Species	Route	Dose	Result	Reference
Mouse	s.c.	25 mg/kg/day (Day 5–15)	Skeletal malformation	13,14
Rat	diet	50–100 ppm (= 5–10 mg/kg/ day)	No effects	15
	drinking water	6 mg/kg/day (Day 0–21)	Reduced number, birth weight & rearing activity (males only) Reduced plasma corticosterone (both sexes)	16
	s.c.	0.5–5 mg/kg/day	Low birthweight Prolonged gestation No abortions	15
	s.c.	0.1 mg/kg/day (from Day 14)	No effects	17
	s.c.	1 mg/kg/day	Reduced litter size Increased stillbirths	17
Man	Inhalation	As in smoking	Low birth weight* No increase in birth defects	

Abbreviations: s.c. = subcutaneously; p.o. = orally; i.p. = intraperitoneally.
*Other smoke components (e.g. CO) possibly responsible.

exposure to nicotine by the inhalation route have been published. In response to dietary nicotine no effects were seen in one study and in response to nicotine given by subcutaneous injection, low birth weight and reduced litter size, but no malformations, were seen at doses in the range of up to 5 mg/kg/day. A report of skeletal malformation in mice given 25 mg/kg/day is uninterpretable because the dosage was so unrealistically high.

Table 4.4 summarizes the available mutagenicity and clastogenicity test data for nicotine. An overview of the results of 85 *in vitro* tests for mutagenicity or clastogenicity on different smoke condensates and condensate fractions led Mizusaki *et al.*[21] to the view that the nicotine content of the materials tested is not positively correlated to the outcome of tests. It is reasonable to conclude that nicotine is neither a mutagen nor a clastogen and that it is not metabolically activated to such by the enzymes present in rat liver microsomes.

Table 4.5 summarizes the reasons why some investigators suspect that

Table 4.4. Evaluation of nicotine and smoke condensates for mutagenicity and clastogenicity

(A) Nicotine and its metabolites

Test	Result	Reference
Ames	Negative	18
Sister-chromatid exchange	Negative	19
DNA-cell-binding assay	Negative	20

(b) Nicotine as a constituent of smoke condensates and condensate fractions

Positivity of results unrelated to nicotine content of material tested in 85 *in vitro* tests of mutagenicity or clastogenicity	21

nitroso-derivatives of nicotine and related alkaloids may, theoretically, be implicated in human lung carcinogenesis. There is, in fact, no direct evidence of such implication. In particular, there are no data from animal studies involving inhalation exposure to any of the tobacco-derived nitrosamines.

Finally, Table 4.6 summarizes reasons for doubting the importance of nicotine as a precursor of carcinogens in tobacco smoke. Many known carcinogens have been detected in trace amounts in tobacco smoke and several of these would be expected to act in the lung at the site of their deposition. Whether this would happen in the case of nicotine-derived nitrosamines is not certain. Certainly, in animals, the oesophagus and nasal cavity are major target sites for the carcinogenicity of these compounds,

Table 4.5. Reasons why some investigators suspect that nitroso-derivatives of nicotine and related alkaloids may be implicated in human lung carcinogenesis

NNN, NNK and NAB are present in raw tobacco and in tobacco smoke.

These compounds are carcinogenic in animals—with the nose and oesophagus as the main target sites.

The lung is one of several targets for carcinogenicity in the case of NNN and NNK

Abbreviations: NNN, *N'*-nitrosonornicotine; NNK, 4-(methylnitrosoamino)-1-(3-pyridyl)-1-butanone; NAB, *N'*-nitrosoanabasine.

Table 4.6. Reasons for doubting importance of nicotine as precursor of carcinogens in tobacco smoke

---

Many carcinogens unrelated to nicotine present in trace concentrations in tobacco and tobacco smoke.

No evidence from inhalation studies that the lung is a *principal* target for NNN or NNK carcinogenicity in animals.

Oesophagus and nasal cavity are major targets for NNN and NNK carcinogenicity in rats and hamsters. These are clearly *not* major risk sites in smokers.

Low incidence of lung cancer in pipe smokers suggests that nitrosamines derived from tobacco alkaloids are not important.

---

whereas in smokers it is clear that these sites are no more than minor target sites. Again, the lowness of the cancer risk in pipe-smokers argues against the notion that nitrosamines derived from tobacco alkaloids are important carcinogens in man.

4.3. Discussion

A serious problem of our time is that chemical analytical methodology has gone streaking ahead of biological and investigative pathological techniques in terms of sensitivity. Thus, we are in a position in which we can detect minute concentrations of substances in tobacco and tobacco smoke which have been shown to be carcinogenic in animals, but cannot ascertain whether, at these very low concentrations, the presence of those substances in smoke constitutes any health risk. This is the present position with regard to nicotine derived nitrosamines. Neither the sheer volume of studies in this area nor the elegance of the chemical analytical work involved should be allowed to influence judgement in this area, for much more harm could be done to society by reducing the ratio of nicotine to other components in smoke than by leaving it alone or increasing it.

Knowledge of factors which cause lung cancer in man is still incomplete and still growing. Early work on tobacco carcinogenesis laid the blame with carcinogenic polycyclic hydrocarbons. Later co-carcinogens in the shape of phenols and other substances became fashionable[22] and later still came the nicotine derived nitrosamines.[23,24] With half the components of smoke as yet unidentified, it would be imprudent to presume that the carcinogenicity of smoke can be explained definitively in terms of what is presently known. It would, therefore, also be imprudent to follow a strategy for safer smoking based on the assumption that nicotine is a significant source of the carcinogenicity of tobacco smoke.

Irrespective of the role of nicotine-derived nitrosamines in tobacco carcinogenesis, it is obviously sensible to reduce the yields of these substances from cigarettes as much as possible. In the UK this has been done primarily by the choice of tobacco which is low in nitrate. Table 4.7 illustrates recent data for NNN and NNK in typical UK cigarette tobacco and in the smoke from cigarettes made of such tobacco.

Table 4.7. Non-volatile nitrosamines in UK cigarette tobacco and in delivered cigarette smoke (unpublished Industry data)

Type of cigarette	Tobacco (ng/cig)		Smoke (ng/cig)	
	NNN	NNK	NNN	NNK
Middle tar, filter unventilated				
Brand 1	343	273	46	41
2	158	236	17	33
3	247	385	21	45
Low tar, filter ventilated				
Brand 4	409	313	42	31
5	261	381	14	30

Abbreviations: see Table 4.5.

References

1. IARC (1986) *Tobacco Smoking*, Monographs on the evaluation of the carcinogenic risk of chemicals to humans, Vol. 38. IARC, Lyon, France.
2. Cohen, A. J. and Roe, F. J. C. (1981). Monograph on the pharmacology and toxicology of nicotine. In *Occasional Paper No. 4*, (ed. A. J. Cohen and F. J. C. Roe), pp. 1–45. Tobacco Advisory Council, London.
3. Patty, F. A. (1963). *Industrial hygiene and toxicology. Vol. II. Toxicology* (ed. D. W. Fassett and D. D. Irish) pp. 2193–2196. Interscience Publishers, New York.
4. Larson, P. S., Haag, H. B., and Silvette, H. (1961). *Tobacco: experimental and clinical studies. A comprehensive account of the world literature.* Williams & Wilkins Co., Baltimore.
5. NIOSH (1977). *Registry of toxic effects of chemical substances*, Vol. 2. National Institution for Occupational Safety and Health, DHEW (NIOSH) Publication No. 78–104–B.
6. Larson, P. S. and Silvette, H. (1971). *Tobacco: experimental and clinical studies. A comprehensive account of the world literature. Supplement II.* Williams & Wilkins Co., Baltimore.
7. Staemmler, M. (1936). Anatomical findings in experimental poisoning with nicotine. *Klinische Wochenschrift*, **15**, 1579.
8. Roe, F. J. C. and Baer, A. (1985). Enzootic and epizootic adrenal medullary

proliferative disease of rats: Influence of dietary factors which affect calcium absorption. *Human Toxicology*, **4**, 27–52.

9. US Surgeon General (1981). Section 2: Pharmacology and Toxicology. In *The health consequences of smoking: the changing cigarette*. US Department of Health and Human Services (PHS) 81–50156, pp. 27–73, US Government Printing Office, Washington DC.

10. Wynder, E. L., Hoffmann, D., and Gori, G. D. (1976). Smoking and Health. In *Proceedings of 3rd World Conference, New York City, June 1975, Volume 1. Modifying the risk for the smoker* (ed. E. L. Wynder, D. Hoffmann, and G. D. Gori), pp. 1–564 DHEW Publication No. (NIH) 76–1221. US Government Printing Office, Washington DC.

11. Roberts, C. J. and Diggle, R. (1982). Non smoking: a feature of ulcerative colitis. *British Medical Journal*, **285**, 440.

12. Perera, D. R., *et al.* (1984). Smoking and ulcerative colitis. *British Medical Journal*, **288**, 1533.

13. Nishimura, H. and Nakai, K. (1958). Developmental abnormalities in offspring of pregnant mice treated with nicotine. *Science, NY*, **12**, 877.

14. DiPaulo, J. A. and Kotin, P. (1966). Teratogenesis-oncogenesis: A study of possible relationships. *Archives of Pathology*, **81**, 3.

15. US Surgeon General (1972). In *The health consequences of smoking*. DHEW Publication No. (HSM) 72–7516, pp. 83–91. US Government Printing Office, Washington DC.

16. Peters, D. A. V. and Tang, S. (1982). Sex-dependent biological changes following prenatal nicotine exposure in the rat. *Pharmacology Biochemistry and Behaviour*, **17**, 1077–82.

17. Hamosh, M., Simon, M. R., and Hamosh, P. (1979). Effect of nicotine on the development of fetal and suckling rats. *Biology of the Neonate*, **35**, 290.

18. Florin, I., Rutberg, L., Curvall, M., and Enzell, C. R. (1980). Screening of tobacco smoke constituents for mutagenicity using the Ames' test. *Toxicology*, **18**, 219–32.

19. Riebe, M. and Westphal, K. (1983). Studies on the induction of sister-chromatid exchanges in chinese hamster ovary cells by various tobacco alkaloids. *Mutation Research*, **124**, 281–6.

20. Kubinski, H., Gutze, G. E., and Kubinski, Z. O. (1980). DNA-Cell-Binding (DCB) assay for suspected carcinogens and mutagens. *Mutation Research*, **89**, 95–136.

21. Mizusaki, S., Akamoto, H., Akiyama, A., and Fukuhari, Y. (1977). Relation between chemical constituents of tobacco and mutagenic activity of cigarette smoke condensate. *Mutation Research*, **48**, 29–36.

22. Roe, F. J. C., Salaman, M. H., and Cohen, J. (1959). Incomplete carcinogens in cigarette smoke condensate: tumor-promotion by a phenolic fraction. *British Journal of Cancer*, **13**, 623–33.

23. Boyland, E., Roe, F. J. C., Gorrod, J. W., and Mitchley, B. C. V. (1964a). The carcinogenicity of nitrosoanabasine, a possible constituent of tobacco smoke. *British Journal of Cancer*, **18**, 265–70.

24. Boyland, E., Roe, F. J. C., and Gorrod, J. W. (1964b). Induction of pulmonary tumours in mice by nitrosonornicotine, a possible constituent of tobacco smoke. *Nature*, **202**, 1126.

II

Smoking Habits and Related Mortality in the UK

5

Trends in cigarette smoking habits in the United Kingdom, 1905–1985

STEPHANIE KIRYLUK and NICHOLAS WALD

Abstract

The consumption of cigarettes per person and the prevalence of cigarette smoking have been falling among men and women since the mid-1970s. Large declines in the sales-weighted average tar and nicotine yields of all brands of cigarettes occurred during the 1970s. Between 1980 and 1985 the sales-weighted average tar yield continued to decline, while there was virtually no change in the sales-weighted nicotine yield. The decline in the sales-weighted average tar yield since 1972 was due to the decrease in the tar yields of existing brands as much as to the introduction of low tar brands. The switch to low tar cigarettes has plateaued in recent years. Low tar cigarettes are more popular with women than with men, with older than with younger people, and among those with professional rather than manual occupations.

The data described in this paper are included in UK Smoking Statistics.[1] *The original source of the data is given at the foot of each table or figure, while complete references to and descriptions of these sources can be found in Wald* et al.[1] *There are some small differences between sets of prevalence data, depending on their source.*

5.1. Prevalence and consumption trends

Per person consumption of cigarettes, 1905–1985

Cigarette* consumption among men rose steadily from about 1890 to reach a peak of 12 cigarettes a day per adult male in 1945, and remained fairly stable, varying between 9 and 11 cigarettes a day for the next 30 years (Fig. 5.1). Since 1974 the per person consumption among men has declined steadily to about 6.5 cigarettes a day by 1985.

Cigarette smoking among women began about 1920, rose to a maximum of just over 7 cigarettes per adult female in 1974, and declined to just over 5 cigarettes a day by 1985.

*Unless otherwise indicated the term 'cigarette' will mean 'manufactured cigarette'.

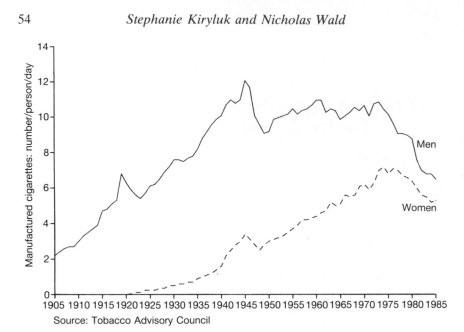

Fig. 5.1. Per person daily consumption of manufactured cigarettes, 1905–1985 (men and women aged 15 and over, UK).

Per person consumption of hand-rolled cigarettes rose from 0.1 cigarette a day per adult in 1931 to a maximum of 0.8 cigarette a day in the mid-1960s; in 1985 it stood at 0.6 cigarette a day per adult. The consumption of hand-rolled cigarettes has not been estimated for men and women separately. Most hand-rolled cigarettes are, however, smoked by men.

The proportion of all tobacco products that are consumed by men as cigarettes has remained at around 80 per cent since 1945 (Fig. 5.2). Very few women smoke tobacco products other than cigarettes.

Prevalence of cigarette smoking, 1948–1985

The percentage of men who smoke cigarettes fell from 65 to 35 per cent between 1948 and 1985 (Fig. 5.3). The percentage of men who smoke hand-rolled cigarettes stood at 13 per cent in 1961, and 11 per cent in 1985. Many of these men also smoke cigarettes; just under 4 per cent smoked hand-rolled cigarettes only in 1985.

The percentage of women who smoke cigarettes fluctuated between 36 and 44 per cent during the years from 1948 to 1975. Since 1975 it has decreased steadily to 34 per cent in 1985, approximately the same level as men.

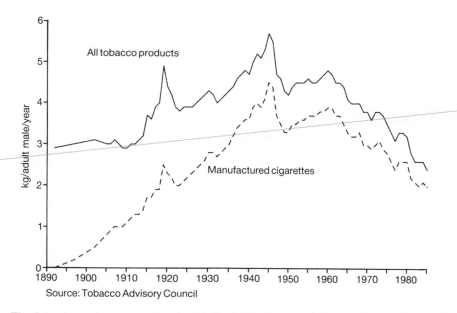

Fig. 5.2. Annual consumption (weight) of all tobacco products and manufactured cigarettes per adult male, 1890–1985 (men and women aged 15 and over, UK).

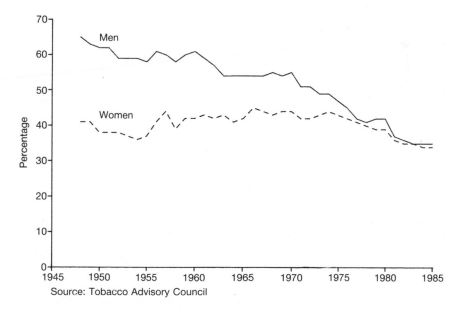

Fig. 5.3. Prevalence of manufactured cigarette smoking, 1948–1985 (men and women aged 16 and over, Great Britain).

Daily consumption of cigarettes per cigarette smoker, 1949–1985

The consumption of cigarettes per adult male smoker rose from 14 a day in 1949 to 19 a day in 1955, and remained at about this level until 1970, when there was a small increase to 22 a day by 1973 (Fig. 5.4). After 1980 consumption began to fall again, and in 1985 daily cigarette consumption among male cigarette smokers stood at 19 cigarettes a day. Both the rise in consumption per smoker between 1970 and 1973, and the relative stability in consumption per smoker between 1973 and 1980 could mask several different trends, such as different proportions of light and heavy smokers giving up at different times, some heavy smokers decreasing their consumption on the way to giving up, and possibly some smokers increasing their consumption [see next sub-section].

Consumption of cigarettes per adult female smoker rose steadily from 7 cigarettes a day in 1949 to a maximum of 17 a day in 1976. By 1984 consumption among female cigarette smokers had fallen to 15 cigarettes a day.

Smokers, ex-smokers, never smoked (cigarettes), 1972–1984

Data from the General Household Survey[3] (Table 5.1) show between 1972 and 1984 an increase in the percentage of men who have never smoked

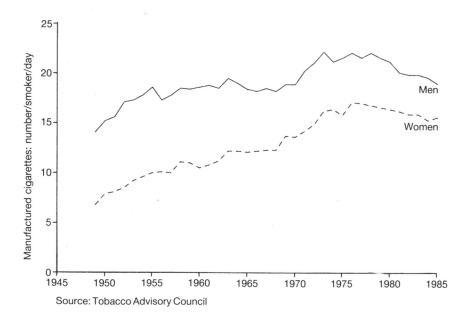

Source: Tobacco Advisory Council

Fig. 5.4. Daily consumption of manufactured cigarettes per manufactured cigarette smoker, 1949–1985 (manufactured cigarette smokers, men and women aged 16 and over, Great Britain).

Table 5.1. Percentages of men and women by cigarette smoking status, 1972 and 1984 (Great Britain)

	< 20 Cigarettes per day	≥20 Cigarettes per day	All cigarette smokers	Ex-smokers of cigarettes	Never smoked cigarettes
Men					
1972	28	24	52	23	25
1984	20	16	36	30	34
Women					
1972	30	11	41	10	49
1984	22	10	32	17	51

Source: General Household Survey.[3]

cigarettes, increases in the percentages of men and women who are ex-smokers, and little change in the percentage of women who have never smoked cigarettes. The percentage of men who were either light (<20 cigarettes a day) or heavy smokers (≥20 cigarettes/day) declined between 1972 and 1984; the decline was most marked among the heavy smokers between 1972 and 1976, and among the light smokers after 1978. Among women the percentage of light smokers declined, with little change in the percentage of heavy smokers.

Prevalence of smoking in young people

There were substantial falls between 1972 and 1984 in the percentages of men and women aged 16–24 who smoke cigarettes (Table 5.2). No such decline, however, is evident among school children. The 1984 OPCS

Table 5.2. Percentages of men and women aged 16–19 years and 20–24 years who smoke cigarettes (Great Britain)

Year	16–19 years		20–24 years	
	Men	Women	Men	Women
1972	43	39	55	48
1974	42	38	52	44
1976	39	34	47	45
1978	35	33	45	43
1980	32	32	44	40
1982	31	30	41	40
1984	29	32	40	36

Source: General Household Survey.[3]

survey of school children[4] estimated that 29 per cent of boys and 30 per cent of girls aged 16 years smoked cigarettes regularly. A number of surveys of smoking habits among school children have been carried out since 1965; the prevalence of smoking among boys appears to have remained fairly constant since the mid-1960s, while smoking among girls has increased to reach, or slightly exceed, that among boys.

Prevalence of cigarette smoking by social class

In 1958 there was little variation across social classes in the percentages of men and women who smoked cigarettes (Fig. 5.5). By 1985 the percentage of men who smoke cigarettes ranged from 23 per cent in social class I (professional) to 43 per cent in social class V (unskilled), while the percentage of women who smoke cigarettes ranged from 26 per cent in social class I to 40 per cent in social classes V and VI.

In 1985 about 17 per cent of men in social classes IV + V + VI were estimated to smoke hand-rolled cigarettes, compared to 5 per cent in social classes I + II. Some of these men also smoke cigarettes or other forms of tobacco.

5.2. Type of cigarette smoked and sales-weighted average yields, 1934–1985

The switch from plain to filter cigarettes began in the 1950s. The original filters were unventilated, and these still account for the majority of the

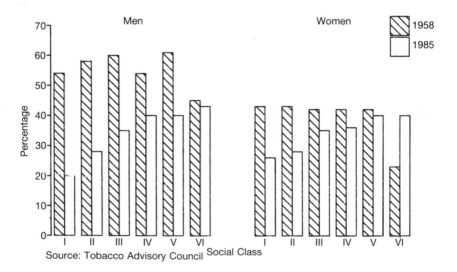

Fig. 5.5. Prevalence of manufactured cigarette smoking: by social class, 1958 and 1985 (men and women aged 16 and over, Great Britain).

filter market. During the 1970s ventilated filters were introduced. By 1985 the market share held by plain, unventilated filter and ventilated filter cigarettes were estimated to be 4, 59, and 37 per cent, respectively.

Sales-weighted average tar yield

Between 1934 and 1985 the sales-weighted average tar yield of all brands of manufactured cigarettes declined, slowly at first, and more rapidly after 1970, from 33 to 14 mg/cigarette. At any point in time plain cigarettes have had a higher sales-weighted average tar yield than filter cigarettes, and unventilated filters have had a higher yield than ventilated filters (Table 5.3, Figs 5.6 and 5.7).

Sales-weighted average nicotine yield

The sales-weighted average nicotine yield of all cigarettes rose from 2.0

Table 5.3. Annual sales-weighted tar yield (mg/cigarette): by type of manufactured cigarette, 1934–1985 (UK)

Year	Plain	Unventilated filter	Ventilated filter	All brands
1934–40	32.9			32.9
1941–47	32.2			32.2
1948–54	29.5			29.5
1955–61	30.4			30.4
1962–68	29.0	24.0		26.0
1969	29.0	22.0		25.7
1970	29.4	21.2		23.6
1971	28.5	20.6		22.3
1972	28.0	19.5	12.0	21.0
1973	26.5	18.9	11.0	19.9
1974	25.4	18.8	9.4	19.3
1975	25.3	18.5	9.3	18.8
1976	24.7	18.0	9.3	17.9
1977	24.6	18.2	9.8	17.6
1978	22.3	18.3	9.8	17.5
1979	18.7	17.9	9.7	16.8
1980	18.2	17.7	9.5	16.4
1981	18.3	16.9	9.3	15.9
1982	17.8	16.5	9.6	15.4
1983	17.5	16.2	10.7	15.1
1984	17.6	15.8	11.9	14.8
1985	17.1	15.6	12.1	14.4

Source: 1934–1940 to 1979 from Wald, Doll and Copeland,[5] based on analysis by the Laboratory of the Government Chemist (LGC). 1980–1985 calculated from LGC surveys together with market share data from *Tobacco*.[7]

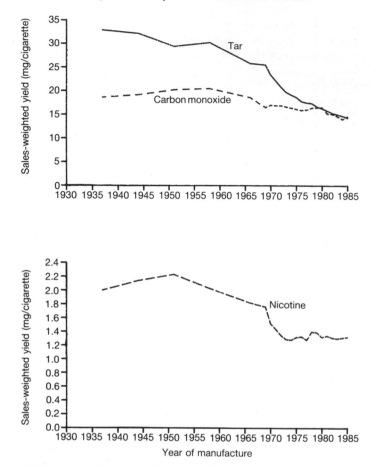

Fig. 5.6. Sales-weighted tar, nicotine and carbon monoxide yields of manufactured
cigarettes (all brands), 1934–1985 (UK). Source: Tables 5.3–5.5.

mg/cigarette in 1934 to 2.2 mg/cigarette in about 1950, and fell to 1.3 mg/
cigarette by 1980. Between 1980 and 1985 there was little change in the
sales-weighted average nicotine yield of all brands of cigarette. In 1985
both plain and unventilated cigarettes had a sales-weighted average nicotine yield of about 1.4 mg/cigarette, while that of ventilated filter cigarettes
was just under 1.2 mg/cigarette (Table 5.4, Figs 5.6 and 5.7).

Sales-weighted average carbon monoxide yield

The sales-weighted average carbon monoxide yield of all brands of cigarette
rose from 18.6 mg/cigarette in 1934 to 20.6 mg/cigarette in about 1958;
thereafter, it declined slowly until about 1970 and then more rapidly, to

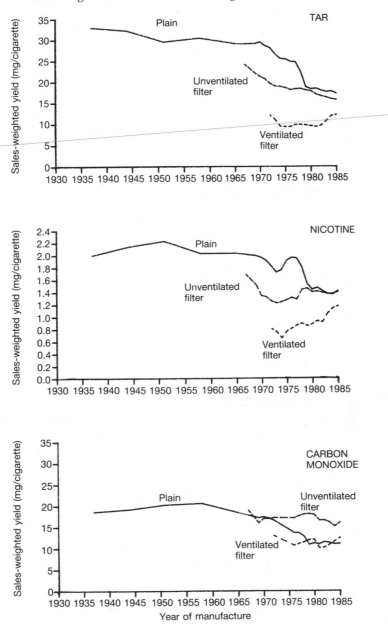

Fig. 5.7. Sales-weighted tar, nicotine and carbon monoxide yields of manufactured cigarettes: by type of cigarette, 1934–1985 (UK). Source: Tables 5.3–5.5.

Table 5.4. Annual sales-weighted nicotine yield (mg/cigarette): by type of manufactured cigarette, 1934–1985 (UK)

Year	Plain	Unventilated filter	Ventilated filter	All brands
1934–40	2.00			2.00
1941–47	2.14			2.14
1948–54	2.23			2.23
1955–61	2.03			2.03
1962–68	2.03	1.68		1.82
1969	1.99	1.51		1.76
1970	1.96	1.34		1.52
1971	1.91	1.32		1.44
1972	1.82	1.25	0.80	1.35
1973	1.72	1.22	0.75	1.29
1974	1.76	1.24	0.65	1.28
1975	1.91	1.27	0.76	1.32
1976	1.96	1.31	0.81	1.33
1977	1.95	1.28	0.86	1.28
1978	1.81	1.44	0.89	1.40
1979	1.54	1.46	0.85	1.39
1980	1.44	1.40	0.88	1.32
1981	1.47	1.41	0.93	1.34
1982	1.42	1.39	0.91	1.31
1983	1.37	1.37	1.05	1.30
1984	1.36	1.37	1.14	1.31
1985	1.42	1.40	1.17	1.32

Source: 1934–1940 to 1979 from Wald, Doll and Copeland,[5] based on analysis by the Laboratory of the Government Chemist (LGC). 1980–1985 calculated from LGC surveys together with market share data from *Tobacco*.[7]

14.7 mg/cigarette in 1985. Plain and ventilated filter cigarettes have lower sales-weighted average carbon monoxide yields than cigarettes with unventilated filters (Table 5.5, Figs 5.6 and 5.7).

5.3. Low tar smoking and the UK cigarette market, 1972–1985

The decline in the sales-weighted average tar yield

The Laboratory of the Government Chemist surveys[6] of the tar, nicotine, and later carbon monoxide, yields of manufactured cigarettes began in 1972. Since then the number of brands on the market have increased. Many, but not all, of the new brands have low tar yields. In addition, there has been a downward shift in the tar yields of many existing cigarette brands, while some have disappeared from the market altogether (Table

Table 5.5. Annual sales-weighted carbon monoxide yield (mg/cigarette): by type of manufactured cigarette, 1934–1985 (UK)

Year	Plain	Unventilated filter	Ventilated filter	All brands
1934–40	18.6			18.6
1941–47	19.2			19.2
1948–54	20.3			20.3
1955–61	20.6			20.6
1962–68	18.4	18.9		18.7
1969	17.2	16.0		16.6
1970	17.4	17.0		17.1
1971	17.1	17.0		17.0
1972	16.6	17.2	13.0	17.0
1973/75	—	—	—	—
1976	13.7	17.1	10.7	16.0
1977	13.6	17.6	11.1	16.1
1978	12.8	18.0	11.5	16.5
1979	10.9	18.1	12.0	16.6
1980	11.2	17.9	12.1	16.6
1981	11.0	16.6	10.1	15.3
1982	11.5	16.6	10.2	15.2
1983	11.3	16.1	10.9	14.7
1984	11.0	15.2	11.7	14.1
1985	11.1	16.1	12.5	14.7

Source: 1934–1940 to 1979 from Wald, Doll and Copeland,[5] based on analysis by the Laboratory of the Government Chemist (LGC). 1980–1985 calculated from LGC surveys together with market share data from *Tobacco*.[7]

5.6). Thus, the decline in the sales-weighted average tar yield since 1972 is due to the decrease in tar yields of existing brands as much as to the introduction of low tar cigarettes.

Most popular cigarette brands: fragmentation of the cigarette market

The leading cigarette brands that together comprised just under 60 per cent of the total market share of manufactured cigarettes in 1972 and 1985 are listed in Table 5.7. This proportion of the market was covered by only six brands in 1972, but by ten brands in 1985. The remainder of the market was taken up by a large number of other brands, some of them occupying 0.1 per cent or less of the total market share. The cigarette market was thus more fragmented in 1985 than in 1972, with the leading brands tending to occupy a lower overall market share. In 1985 the leading brand had a middle tar yield (16 mg/cigarette), and only one low tar cigarette appeared in the 10 most popular brands.

Table 5.6. Number of brands tested in each tar band of the Laboratory of the Government Chemist surveys, 1, 21, and 23 (1972, 1984, 1985–1986)

	Tar yield (mg/cigarette) before 1985					Total number of brands tested
	< 10 low	11–16 low/middle	17–22 middle	23–28 middle/high	⩾ 29 high	
Survey 1 July 1972– December 1972	5	19	53	13	11	101
Survey 21 July 1984– December 1984	33	72	30	2	0	137

	Tar yield (mg/cigarette) after 1985				
	0–9.99 low	10.0–14.99 low/middle	15.0–17.99 middle	18 + high	
Survey 23 September 1985– February 1986	33	53	42	10	138

Source: Laboratory of the Government Chemist.[6]

Low tar smoking

In the United Kingdom low tar cigarettes were defined as having tar yields of less than 10.5 mg/cigarette for the years 1972–1984, and less than 10 mg/cigarette from 1985 onwards.

Low-tar smoking reached a plateau in the late 1970s. Data from National Opinion Polls Market Research Limited (NOP)[9] show that the percentage of all cigarette smokers (including hand-rolled smokers) who smoke low tar cigarettes rose to just under 10 per cent of men and 20 per cent of women by 1977; since then there has been little change (Fig. 5.8). Figure 5.9 displays data from a 1985 NOP survey that categorizes all cigarette smokers by the tar yield of the brand of cigarette that they usually smoke. Cigarettes in the middle tar band (15.00–17.99 mg tar/cigarette) are smoked by the largest proportion of both male and female smokers. (Less than 0.5 per cent of respondents interviewed smoked cigarettes in the high tar category.)

In 1984 the General Household Survey estimated that 11 per cent of male and 23 per cent of female manufactured cigarette smokers were smoking low tar cigarettes. The survey showed that the percentage of low tar smokers rose markedly with age in women, but less so with men. The percentage of low tar smokers was lowest among those aged 16–19 (6 per

Table 5.7. Most popular cigarette brands, 1972 and 1985

	Market share (%)	Tar	Nicotine	Carbon Monoxide
			Yields (mg/cigarette)	
1972				
Players No 6 Filter	19.2	20	1.2	—
Embassy Filter	19.0	20	1.3	—
Embassy Regal	6.2	18	1.2	—
Park Drive Plain	4.9	28	1.9	—
Players No 6 Filter	4.6	20	1.3	—
Benson & Hedges KS	4.6	20	1.3	—
Total market share	58.5			
1985				
Benson & Hedges Special Filter KS	18.2	16	1.5	17
John Player Superkings	6.3	13	1.4	14
Embassy Regal KS	6.1	13	1.3	15
John Player Special KS	6.0	16	1.4	16
Silk Cut KS	5.6	9	0.9	9
Embassy No 1 KS	3.1	13	1.2	15
Rothmans KS	2.7	16	1.6	15
Berkeley LL Superkings	2.7	14	1.4	14
Raffles	2.6	14	1.4	16
Dunhill KS	2.6	14	1.3	15
Total market share	55.9			

Source: Market share for 1972 from *World Tobacco*,[8] for 1985 from *Tobacco*.[7] Yields from surveys 1 and 23 of the Laboratory of the Government Chemist; carbon monoxide was not analysed in 1972. (NB yield data were rounded in 1972 and truncated in 1985.)

cent of men and 10 per cent of women), although the tar yield could not be determined for about 20 per cent of this age group. The age group 16–24 years reported the highest prevalence of middle/high tar smoking (17–22 mg tar/cigarette – the highest yield that was tabulated). This trend towards increasing popularity of low tar cigarettes with increasing age is also evident in 1985 data from NOP (Fig. 5.10).

A social class gradient has always been evident in the percentage of low tar smokers. A 1985 NOP survey showed that the prevalence of low tar smoking among men and women in the higher and intermediate professional classes was more than twice that among men and women who were manual workers (Fig. 5.11).

Data on percentage market share are not available from the tobacco industry, but estimates of the percentage of sales in the low tar band can be made from the tobacco trade journals.[7,8] Low tar sales rose steadily from

Stephanie Kiryluk and Nicholas Wald

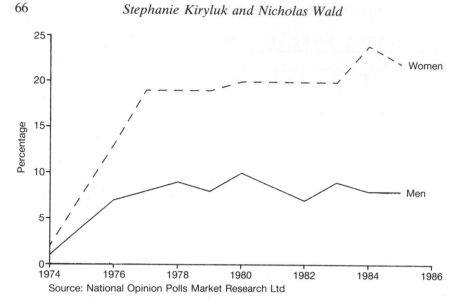

Source: National Opinion Polls Market Research Ltd

Fig. 5.8. Percentages of cigarette smokers (including hand-rolled smokers) who smoke low tar cigarettes, 1974–1985 (cigarette smokers, men and women aged 16 and over, Great Britain).

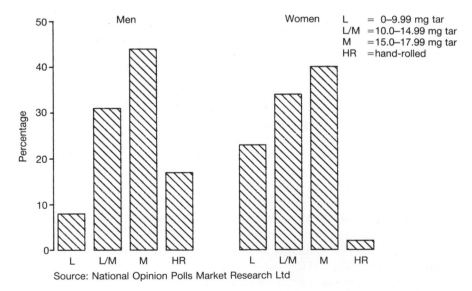

Source: National Opinion Polls Market Research Ltd

Fig. 5.9. Percentages of cigarette smokers: by tar yield of cigarette smoked, 1985 (cigarette smokers, men and women aged 16 and over, Great Britain).

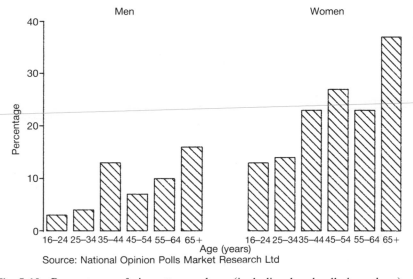

Men   Women

Source: National Opinion Polls Market Research Ltd

Fig. 5.10. Percentages of cigarette smokers (including hand-rolled smokers) who smoke low tar cigarettes: by age, 1985 (cigarette smokers, men and women aged 16 and over, Great Britain).

1972 until they accounted for about 12 per cent of the cigarette market in 1977. Between 1977 and 1985 low tar sales fluctuated between 12 and 15 per cent of the market, and stood at about 14 per cent in 1985.

Changes in the UK cigarette market in 1985

According to the trade journal *Tobacco*[7] one of the main areas of growth in market share in 1985 were the private labels, including supermarket 'own brands'. Many of these are imported; some are low tar, but many have fairly high tar yields. Another area of growth has been some low tar brands including Rothmans Special, which has a low tar and relatively high nicotine yield (9 mg tar and 1.3 mg nicotine per cigarette in 1985).

5.4.   Effect of price on smoking habits

Experience has shown that taxation can be used to influence the types of cigarette that are sold. King size cigarettes became popular in the United Kingdom when taxation changed to an amount per cigarette rather than according to weight. In 1985 some longer length cigarettes, possibly perceived as being better value for money, increased their market share.

Stephanie Kiryluk and Nicholas Wald

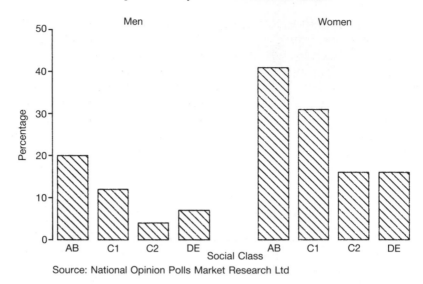

Source: National Opinion Polls Market Research Ltd

Fig. 5.11. Percentages of cigarette smokers (including hand-rolled smokers) who smoke low tar cigarettes: by social class, 1985 (cigarette smokers, men and women aged 16 and over, Great Britain).

Social Class A: higher managerial, administrative or professional.
Social Class B: intermediate managerial, administrative or professional.
Social Class C1: supervisory or clerical, and junior managerial, administrative or professional.
Social Class C2: skilled manual workers.
Social Class D: semi- and unskilled manual workers.
Social Class E: state pensioners or widows, casual or lowest grade workers, long-term unemployed).

Smokers could be encouraged to smoke low tar cigarettes by taxing such cigarettes at a lower rate than other cigarettes; any change in taxation policy must be agreed by the members of the European Community.

5.5. Conclusions

Since about 1970 there have been substantial falls in the prevalence of cigarette smoking and in the *per capita* consumption of cigarettes among men and women. Nevertheless, in 1985 some 35 per cent of men and women still smoked cigarettes. While the sales-weighted average tar yield continues to decline, there has been little change in the sales-weighted average nicotine yield of all brands of cigarettes between 1980 and 1985. The uptake of low tar smoking, measured in terms of both percentage of smokers who smoke low tar cigarettes, and in terms of market share reached a plateau in the late 1970s, with little change since then. Particularly in recent years, therefore, the decline in the sales-weighted average tar yield has been due to the decrease in the tar yield of existing brands, as much as to a switch to low tar smoking. Low tar cigarettes are smoked predominantly by certain sectors of the population; they are smoked more by women than men, more by the elderly than the young, and more by professional than manual workers. Any future modifications to cigarettes that do not apply to the entire cigarette market may need to be accompanied by other measures such as differential taxation to encourage their uptake throughout the population of those who continue to smoke.

References

1. Wald, N., Kiryluk, S., Darby, S., Doll, R., Pike, M., and Peto R. (1988). *UK Smoking Statistics*. Oxford University Press, Oxford.
2. Lee, P. N. (ed.) (1976). *Statistics of smoking in the United Kingdom. (Research Paper 1) 7th edition*. Tobacco Research Council, London (Tobacco Advisory Council data for 1976 onwards are to be found in Wald *et al.*).
3. Office of Population Censuses and Surveys. *General Household Survey 1972*: 1975, *General Household Survey 1974*: 1977, *General Household Survey 1976*: 1978, *General Household Survey 1978*: 1980, *General Household Survey 1980*: 1982, *General Household Survey 1982*: 1984, *General Household Survey 1984*: 1986. HMSO, London.
4. Dobbs, J. and Marsh, A. (1985). *Smoking among secondary school children in 1984*. HMSO, London.
5. Wald, N., Doll, R., and Copeland, G. (1981). Trends in the tar, nicotine and carbon monoxide yields of UK cigarettes manufactured since 1934. *British Medical Journal*, **282**, 763–5.
6. *Tar and Nicotine Yields of Cigarettes. Tar, Carbon Monoxide and Nicotine Yields of Cigarettes*. Leaflets issued by the Health Departments of the United Kingdom.
7. Anonymous. *Tobacco*, August 1981: 24–25, August 1982: 18–19, September 1983: 20–1, September 1984: 7, September 1985: 4–5, August 1986: 7–9. Brand Shares of the UK cigarette market.
8. Maxwell, J.C. (1973). How the Brands Ranked. Maxwell International Estimates. *World Tobacco*, 41:54.
9. National Opinion Polls Market Research Limited. *Smoking Habits*. Unpublished reports carried out by National Opinion Polls Market Research Limited for the Office of Population Censuses and Surveys.

6

Trends in mortality from smoking-related diseases in England and Wales

SARAH C. DARBY, RICHARD DOLL, and IRENE
M. STRATTON

Abstract

Data are presented on the recent trends in mortality in England and Wales from lung cancer, ischaemic heart disease, aortic aneurysm, and chronic obstructive lung disease, considered to be the main smoking-related diseases. For the most part the trends can only be explained by changes in the real incidence of the diseases. Furthermore, any changes in the level of disease attributable to manufactured cigarettes must be due to changes in the constituents of cigarettes rather than the number smoked. However, the possibility of distinguishing the effects of alterations in the level of the main cigarette components, tar, nicotine, and carbon monoxide, purely from national trend data seems remote.

6.1. Introduction

The objective of this paper is to present data on the recent trends in mortality from the main smoking-related diseases in England and Wales, and to discuss briefly whether or not the observed trends represent real changes in the level of disease, or whether they are due to changes in the efficacy of treatment, the accuracy of diagnosis, or the terminology used to describe and classify the diseases. A fuller report will be published elsewhere,[1] but this paper should provide a basis for examining the extent to which the real trends in these diseases are due to variation in the amounts and types of tobacco products smoked, or to other agents.

6.2. Smoking related diseases

The recent International Agency for Research on Cancer (IARC) monograph[2] evaluating the carcinogenic risk of tobacco smoking to humans has grouped the various causes of death into five categories according to the way in which they are related to smoking, as shown in Table 6.1. Category A includes six diseases where the evidence suggests that practically the whole of the difference in mortality between smokers and life-long non-

Table 6.1. Importance of different causes of death related to smoking in different ways in England and Wales[1]

Category*	Cause of death	No. of deaths as % of total deaths in England & Wales, 1984
A	Cancer of lung	6.3
	Ischaemic heart disease	27.8
	Respiratory heart disease	0.4
	Aortic aneurysm	1.3
	Peripheral vascular disease	0.3
	Chronic obstructive lung disease	4.3
	Subtotal	40.4
B	Alcoholism	<0.1
	Cirrhosis of liver	0.4
	Poisoning	0.2
	Suicide	0.8
	Subtotal	1.4
C	Cancer of oesophagus	0.8
	Cancer of lip, tongue, mouth, pharynx, larynx	0.4
	Cancer of stomach	1.8
	Cancer of liver	0.2
	Cancer of bladder	0.8
	Cancer of kidney	0.4
	Cancer of pancreas	1.1
	Cancer of cervix uteri	0.3
	Cancer of unspecified site	1.6
	Respiratory tuberculosis	<0.1
	Pneumonia	4.4
	Other respiratory disease	1.4
	Myocardial degeneration	0.9
	Hypertension	0.8
	Arteriosclerosis	1.2
	Cerebral thrombosis	2.5
	Other cerebrovascular disease	10.1
	Peptic ulcer	0.8
	Hernia	0.2
	Osteoporosis	0.2
	Subtotal	29.9
D	Cancer of endometrium	0.2
	Parkinsonism	0.6
	Ulcerative colitis	<0.1
	Toxaemia of pregnancy	<0.1
	Subtotal	0.8
E	All others	27.5

*A, diseases for which excess mortality in smokers is attributable to smoking; B, diseases for which excess mortality in smokers is attributable to confounding; C, diseases for which excess mortality in smokers may be partly or wholly attributable to smoking; D, diseases for which excess mortality in non-smokers may be preventable by smoking; E, other diseases.

smokers is due to tobacco. Four of these: cancer of the lung, ischaemic heart disease, aortic aneurysm, and chronic obstructive lung disease, have been selected for the present study. Between them these four accounted for about 40 per cent of the total deaths registered in England and Wales in 1984. Respiratory heart disease has been omitted on account of the difficulty of obtaining reliable statistics relating to it, and peripheral vascular disease has been omitted because of its low case-fatality rate, rendering it unsuitable for study using death certificate data. Category C in the IARC monograph includes those diseases that have been positively associated with smoking, and for which the excess mortality in smokers may be partly or wholly attributable to smoking. Some of these are cancers of specific sites, and the available human data are reviewed in the monograph. For eight sites, namely oesophagus, lip, oral cavity (excluding the salivary gland), pharynx (excluding the nasopharynx), larynx, bladder, kidney, and pancreas, the monograph concludes that there is sufficient evidence to establish tobacco smoking as an important cause*. Between them these eight sites of cancer accounted for just over 3 per cent of the total deaths registered in England and Wales in 1984. Data for them will be presented elsewhere.[1] Myocardial degeneration is included in IARC Category C, and we have included myocardial degeneration with reference to arteriosclerosis in our definition of ischaemic heart disease prior to 1968 because of the frequency with which it was used in the past to describe cases now attributed to the latter. The remaining non-neoplastic diseases in Category C have been omitted from the present study. For peptic ulcer this is because the mortality trends are complex and affected by many factors, including therapy, and for other diseases it is because the relative risks of mortality in smokers compared with non-smokers have been low in most studies of the effects of tobacco smoking. The three remaining disease categories defined by the IARC are: category B, diseases for which excess mortality in smokers is attributable to confounding; category D, diseases for which there is an excess mortality in non-smokers which may be preventable by smoking; and category E, diseases that are generally unrelated to smoking. All these have been excluded, except in so far as the trends in them are relevant to the understanding of the trends in the diseases related to smoking.

6.3.   Data on smoking-related diseases

Annual death rates for the selected diseases have been calculated in 5-year age-groups for males and females between 1950 and 1984 inclusive and will

*In the case of renal cancer, the evidence is sufficient only in relation to the renal pelvis. Adenocarcinoma of the kidney arising from the cortex was classed as 'perhaps' caused by smoking.

be published in full elsewhere.[1] Data prior to 1950 have not been considered because, for many diseases, the diagnosis becomes increasingly uncertain in earlier time periods, and the difficulties in bridging between International Classification of Disease (ICD) codes also increase. Since the early 1950s diagnostic accuracy is thought to have been reasonably reliable for most important diseases, except in the very elderly. For the 85+ age group there is the additional problem that only an aggregated population estimate is published, and changes in the age-distribution of the population within the age group could create distortions in the rates and give rise to misleading trends.

Trends in the annual death rates for each of the selected diseases are plotted in Figs 6.1–6.4 for age groups 30–34, 40–44, 50–54, 60–64, 70–74,

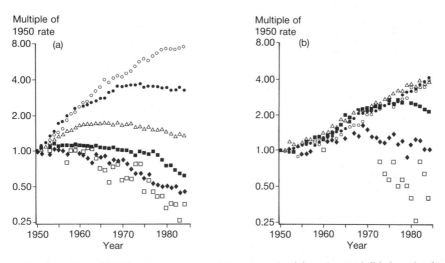

Fig. 6.1. Trends in lung cancer mortality rates in (a) males and (b) females in England and Wales, 1950–1984, at ages 30–34 (□), 40–44 (◆), 50–54 (■), 60–64 (△), 70–74 (●), 80–84 (○). Annual data except as indicated below.

	Age group	Years combined	Plotted in
Males	30–34	1950–56	1953
	80–84	1950–52	1951
Females	30–34	1950–71	1960
	40–44	1950–54	1952
	50–54	1950–52	1951
	60–64	1950–51	1950
	70–74	1950–51	1950
	80–84	1950–53	1951

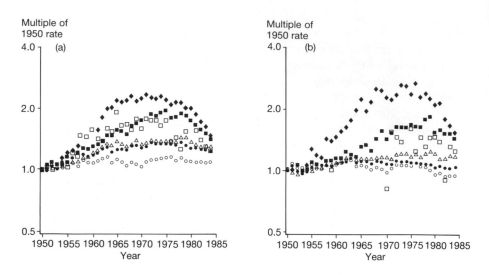

Fig. 6.2. Trends in mortality rates for IHD in (a) males and (b) females in England and Wales, 1950–1984, at ages 30–34 (□), 40–44, (◆), 50–54 (■), 60–64 (△), 70–74 (●), 80–84 (○). Annual data except as indicated below.

	Age group	Years combined	Plotted in
Males	30–34	1950–54	1952
Females	30–34	1950–69	1959
	40–44	1950–53	1951

and 80–84 years. The ICD codes used to define each disease are indicated in Table 6.2. In the present study interest centres on the trend within each age-group, rather than the relative positions of the different age groups, and so the age-specific rates have all been divided by the corresponding rate in 1950, thus enabling trends for a wide range of ages to be plotted on a single figure. In cases where the rate in 1950 was very low and, therefore, subject to variability the divisor has been calculated from the rate for a period spanning several years, starting in 1950, and continuing until the number of deaths was at least 400; that is, sufficient to reduce the standard error of the rate to less than about 5 per cent. The age groups for which this occurred are indicated in footnotes to each figure. For some diseases and sexes, the total numbers of deaths in the entire period was less than 400, and so these age groups have been omitted from the figures for the relevant diseases and sexes.

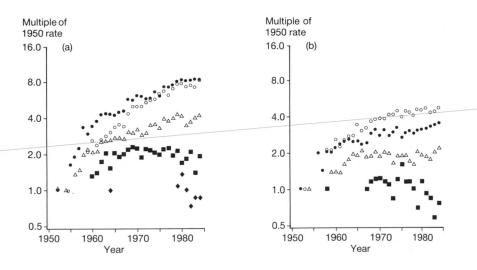

Fig. 6.3. Trends in aortic aneurysm mortality rates in (a) males and (b) females in England and Wales, 1950–1984, at ages 40–44 (◆) (males only), 50–54 (■), 60–65 (△), 70–74 (●), 80–84 (○). Annual data except as indicated below.

	Age group	Years combined	Plotted in
Males	40–44	1950–78	1964
	50–54	1950–59	1954
	60–64	1950–55	1952
	70–74	1950–54	1952
	80–84	1950–58	1954
Females	50–54	1950–66	1958
	60–64	1950–58	1954
	70–74	1950–55	1952
	80–84	1950–57	1953

6.4. Factors that may produce artificial trends in mortality rates

Apart from real changes in the incidence of the disease in question, four factors that may influence trends in certified mortality rates are:

(1) changes in the efficacy of the treatment available may alter the typical length of survival of individuals with the disease;

(2) sudden breaks in the trends may occur when revisions of the International Classification of Diseases (ICD) are introduced;

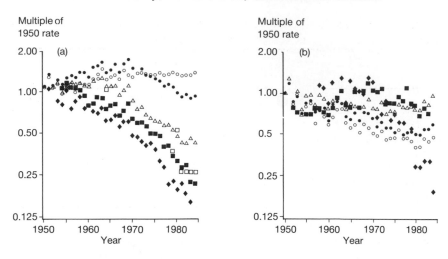

Fig. 6.4. Trends in mortality rates for chronic obstructive lung disease in (a) males and (b) females in England and Wales, 1950–1984, at ages 30–34 (□) (males only), 40–44 (◆), 50–54 (■), 60–64 (△), 70–74 (●), 80–84 (○). Annual data except as indicated below.

	Age group	Years combined	Plotted in
Males	30–34	1950–78	1964
	40–44	1950–52	1951
Females	40–44	1950–57	1953
	50–54	1950–51	1950

(3) changes in the use of medical terms may occur, so that conditions previously categorized under one heading tend to be classed under another;

(4) changes in the accuracy of diagnosis may alter the numbers of people who, dying from the disease, are not recorded as such on the death certificate, or who, dying from some other cause, have their death recorded as due to the disease in question.

The first of these factors reflects an important biological change, the remaining three are human artefacts. In the following sections, the observed trends in the selected diseases will be described, and the role of these four factors will be discussed briefly in relation to them.

Table 6.2. International Classification of Disease (ICD) codes used in calculating the trends in diseases presented in Figs 6.1–6.4

	ICD codes			
	6th revision 1950–57	7th revision 1958–67	8th revision 1968–78	9th revision 1979–84
Cancer of lung	162–3	162–3	162	162
Ischaemic heart disease	420, 422.1	420, 422.1	410–414	410–414
Aortic aneurysm	451	451	441	441
Chronic obstructive lung disease	502, 527	502, 527	491–2, 519	491–2, 496, 519

6.5. Lung cancer

From Fig. 6.1, it appears that the death rates in men in the age groups 30–34, 40–44, and 50–54 years were reasonably stable in the early 1950s while at older ages the death rates were increasing. In age groups 30–34 and 40–44 years the death rate began to decline in the early 1960s. In both age groups the declining trends have continued, so that by 1980–1984 the rates were approximately one-third and one-half, respectively, those in 1950–1954. At older age groups, successively smaller reductions have commenced in successive calendar years, so that by 1984 the death rate was declining in all age groups other than age 80–84 years. At this age the death rate was still increasing so that in 1980–1984 the male rate was over 6 times the male rate in 1950–1954.

The trends in death rates among women are similar to those in men only for age group 30–34 years. In women aged 40–44 and 50–54 years the death rates were increasing during the 1960s. In the 40–44 years age group this was superceded by a decline that started in about 1970 and has continued progressively so that by 1980–1984 the female death rate had returned to its 1950–1954 value. In the 50–54 years age group the death rate did not start to decline until the late 1970s and by 1980–1984 it still remained at about twice the 1950–1954 value. At older ages the female death rates have risen steadily since the early 1950s and do not yet show any sign of a decrease. In these older women the rate of increase in the death rate has been slightly less than that in old men, and by 1980–1984 the female rates were about 3.5 times their 1950–1954 values.

Survival after diagnosis of lung cancer remains poor, with under 10 per cent of patients remaining alive 5 years after diagnosis,[3] so that any changes in the efficacy of the available treatment will have had little impact on trends in mortality. Similarly, there have been no changes in medical nomenclature over the last 35 years which would have affected the lung cancer mortality trends appreciably, and plots of annual lung cancer death rates for males and females in 5-year age groups during the period 1950–1984 show that there is no age group for males or females which shows a sudden break at the time of introduction of a new ICD revision.

The main improvements in recognition of lung cancer happened before the 1950s, and the general level of diagnosis is likely to have been good since then. There remain, however, some specific diseases which could theoretically be a source of diagnostic confusion with lung cancer. For example, the rise in lung cancer death rates at older ages might be due, in part, to a reduction in the numbers of deaths attributed to cancer of unspecified site, pneumonia, or senility, and the reductions in lung cancer death rates in younger males might be partly due to a preferential diagnosis of cancer of the pleura (especially pleural mesothelioma) or cancer of the mediastinum. To examine these possibilities, tabulations of the mortality rates for these diseases have been constructed for males and females for the same time period and age groupings as for lung cancer itself, and the levels of and trends in these diseases compared with those for lung cancer. We have concluded that there is no evident explanation for the recent decrease in lung cancer death rates in males aged 25–54 years other than a real decrease in the incidence of the disease. Nor is there any positive evidence that the continued increases in lung cancer death rates in males aged 80–84 years or females aged 60–84 years are due to anything other than a real increase in the incidence of the disease. However, it is possible that some of the increase in the 80–84 years age group has occurred as a result of a decreasing tendency to certify deaths as due to senility, especially among females.

6.6. Ischaemic heart disease

From Fig. 6.2, male death rates for ischaemic heart disease (IHD) were increasing in the 1950s in all age groups. The rate of increase was greater in men aged under 55 than at older ages, so that by the early 1970s the rates in younger men were approximately double their values in the early 1950s. A turning point occurred, however, during the 1970s, so that by the early 1980s the rate in every age group below 80–84 years was declining. The decline started slightly earlier, and its rate has been somewhat greater, in men aged under 55 than in men aged 60–64 or 70–74 years. However, by 1984 the male death rates had not yet returned to their 1950 level in any age group and in men aged under 55 they were still increased by 50 per cent over their 1950 values.

The trends in female IHD death rates in those aged under 55 are similar to those for males. For women aged 60 and above the rate of increase during the 1950s and 1960s was less than in men; in none of these older age groups has the rate in any year been more than about 30 per cent greater than its 1950 value. In women aged 60–64 years the rate of increase diminished during the 1970s and in the early 1980s the mortality rate was approximately stable, and about 20 per cent higher than its 1950 value. At ages 70–74 years and 80–84 years the female death rates have recently started to decline and are now approximately equal to their 1950 values.

Plots of annual IHD rates for males and females in 5-year age groups during the period 1950–1984 show that there is no age group for males or females which shows a sudden break at the introduction of a new ICD revision. To examine the possibility that diagnostic confusion between IHD and hypertension, or rheumatic and other heart diseases, or a shift in the choice of condition to which death is attributed in patients with both heart disease and pneumonia, has accounted for some of the trends in Fig. 6.2, rates for these other diseases were calculated and compared with those for IHD. There was no evidence to suggest that material changes in ascribing death to pneumonia or IHD has had a major influence on the age-specific trends in IHD rates. In addition, there was no evident explanation for the increase in IHD death rates between the early 1950s and the 1970s in males and females aged 30–54 years, other than real changes in the incidence of, or fatality from IHD. However, the greater availability of coronary care units, greater use of coronary bypass surgery, treatment of hypertension and, in the last few years, better drug therapy, may all have contributed to the decreasing death rates seen in the younger age groups since the 1970s in both males and females.

6.7. Aortic aneurysm

Death rates from aortic aneurysm were increasing in the 1950s in men above age 60 (see Fig. 6.3). Since 1975 the rate of increase has diminished, and by the early 1980s the death rates were relatively stable. In women the trends are broadly similar to those in men. However, in women aged 50–54 years there is some evidence of a decrease in the last few years.

Before the Second World War, syphilis was a major cause of aortic aneurysm. Syphilitic and non-syphilitic aortic aneurysm are clinically distinct, and not likely to be confused diagnostically. However, formerly, syphilitic aortic aneurysm was often referred to simply as aortic aneurysm, whereas more recently cases due to syphilis have become rare, and are more likely to be specified as such. Following the trend in terminology, deaths certified as due to aortic aneurysm, not specified as syphilitic were coded in the sixth revision of the ICD as if they were due to syphilis. After the introduction of the seventh ICD revision in 1958 the coding rules specified that such deaths should be coded as if they were not due to syphilis. To overcome these changes, mortality rates for cardiovascular syphilis have been calculated for the period 1950–1984 and added to those for aortic aneurysm. In men aged 70–74 and 80–84 years, and in women aged 70–74 years the rates for this wider disease group had increased more than three-fold during this period, indicating that mortality rates due to non-syphilitic aortic aneurysm have increased by at least this amount. In women aged 70–74 years there was a two-fold increase, and smaller increases were observed for both sexes at age 60–64 years. Below age 60 it is hard to draw any firm conclusion as the numbers of deaths are so few.

In recent years some cases of aortic aneurysm have been treated surgically before rupture, but they are unlikely to have been on a large enough scale to have had a material impact on the death rate.

6.8.   Chronic obstructive lung disease

In men, mortality rates from chronic obstructive lung disease (COLD) at ages 40–44 and 50–54 years were relatively stable in the early 1950s, see Fig. 6.4. From about 1960 they started to decline. These declines have continued progressively and in 1984 the rates in these age groups were, respectively, 13 and 22 per cent of their 1950 values. In age groups 60–64 and 70–74 years male COLD mortality rates increased in the late 1950s so that in the early 1960s they were increased by 20 and 50 per cent, respectively, compared with their 1950 values. Since then the rates in these two age-groups have declined progressively, and in 1984 were 40 and 90 per cent of their 1950 values. At ages 80–84 years COLD mortality rates in men increased by nearly 40 per cent between 1950 and the late 1970s, and since then have remained relatively stable. In women the trends differ from those in men. At ages 70–74 and 80–84 years the rates have declined progressively since 1950, and in 1984 were approximately one-half their 1950 values. At ages 50–54 and 60–64 years the rates have remained approximately stable and in 1984 were just below their 1950 values. At ages 40–44 years the rate remained approximately equal to its value in the early 1950s until about 1970, but since then has declined steeply, and by the early 1980s was about one-third its value in the early 1950s.

There are no sudden breaks in the annual COLD death rates on introduction of the new ICD revisions. However, there has been a shift from the terms chronic bronchitis and emphysema towards COLD during the period 1950–1984, and so our definition of COLD includes all three. Emphysema was classed with 'other respiratory disease' up to 1968, and COLD was classed with 'other respiratory disease' until 1978. Therefore 'other respiratory disease' has also been included in the definition throughout the whole time period.

Diseases that might be confused diagnostically with COLD are acute and unspecified bronchitis, bronchiectasis, asthma, and pneumonia. However, when trends in mortality from these diseases were compared with those for COLD, no obvious explanation for the large changes in COLD age-specific mortality rates were found, other than real changes in the incidence and severity of the disease. A lower fatality rate could in part have been caused by stronger advice to stop smoking, although changes in treatment will otherwise have made little difference.

6.9. Conclusions

There have been substantial changes in the age-specific mortality rates from many smoking related diseases in England and Wales in the period 1950–1984. For the most part, these cannot be explained in any way other than changes in the real incidence of the disease. The extent to which the changes are due to manufactured cigarettes or to other agents deserves fuller discussion. However, the average weekly consumption by men of manufactured cigarettes at each age group in Britain remained stable during the period 1948–1975,[4] and this provides good evidence that any changes in the level of disease in men at least that are attributable to manufactured cigarettes have been caused by changes in the constituents of the cigarettes, rather than the number smoked on average. As the changes in the tar and nicotine yields of manufactured cigarettes have been quite closely correlated,[5,6] and also somewhat correlated with changes in carbon monoxide levels,[5,6] it is likely to be impossible to distinguish the effects of alterations in the level of any one of these components purely from a study of national trends in smoking related diseases.

Acknowledgements

The trends in mortality rates presented in this paper have been calculated using computerized historic mortality data files made available by the Office of Population Censuses and Surveys.

References

1. Wald, N., Doll, R., Darby, S., Kiryluk, S., Peto, R., and Pike, M. (ed.) (1989) *Trends in smoking related diseases in Britain.* Oxford University Press, Oxford in preparation.
2. IARC (1986). *Tobacco smoking.* IARC monographs on the evaluation of the carcinogenic risk of chemicals to humans, Vol. 38. International Agency for Research on Cancer, Lyon.

3. Cancer Research Campaign (1982). In *Trends in cancer survival in Great Britain*. Cancer Research Campaign, London.
4. Lee, P. N. (ed.) (1976). *Statistics of smoking in the United Kingdom* (7th edn). Tobacco Research Council, London.
5. Wald, N., Doll, R., and Copeland, G. (1981). Trends in tar, nicotine and carbon monoxide yields of U.K. cigarettes manufactured since 1934. *British Medical Journal*, **282**, 763–5.
6. Fairweather, F. A., *et al.* (1981). Changes in the tar, nicotine and carbon monoxide yields of cigarettes sold in the United Kingdom. *Health Trends*, **13**, 77–81.

III
Smoking Yields and Compensation

7

Limitations to potential uses for data based on the machine smoking of cigarettes: cigarette smoke contents

W. S. RICKERT, J. C. ROBINSON, and E. LAWLESS

Abstract

It has been suggested that cigarette smoking could be made less hazardous by reducing the concentration of toxic smoke constituents relative to the concentration of nicotine. This study has shown that the tar/nicotine (T/N) ratio changes significantly in response to changes in smoking conditions. Consequently, it is not clear that the amount of tar per unit nicotine absorbed by smokers would be less if the T/N ratio to smoking machines was decreased. In fact, this study demonstrates that just the opposite might occur, particularly when ultra-low tar cigarettes are smoked. However, even though the T/N ratio can be increased by 50 per cent or more by intense smoking, the maximum T/N ratio for ultra-low cigarettes was still less than that found under any condition for the middle tar cigarettes which were tested.

Studies of smoke absorption in volunteer subjects have demonstrated that published cigarette yields are poor predictors of smoke absorption. Results reported here from a non-volunteer random selection of smokers support this conclusion for long-term users of a particular brand. With respect to switchers, a 12 per cent decrease in smoke absorption was found for those who voluntarily switched to a low tar cigarette in contrast to those who have always smoked this type of brand.

Figures for smoke constituents are sometimes misused by investigators and often misinterpreted by smokers. Based on the questionnaire responses reported here, smokers rate high tar (18 mg) brands as about twice as hazardous as low tar (3 mg) brands. In this survey, 32 per cent of smokers reported switching to a low-tar brand based on this belief. Part of the explanation for a quantitative association of tar yield with risk to health in the smokers mind may lie in their perception of the numbers for tar and nicotine which appear on every package of Canadian cigarettes. Fifty-one per cent of those who responded felt that these numbers represent the most that can be inhaled from a cigarette.

7.1. Introduction

Inhalation of tobacco smoke increases the relative risk for many pathological conditions depending on amount smoked. Those who smoke fewer cigarettes per day have a reduced relative risk with the maximum benefit accruing to those who stop smoking altogether.[1] Since cigarette smoking and drug dependence have many features in common,[2] there are numerous individuals who cannot quit and will continue to smoke in spite of the consequences. It is for this reason that governments or government agencies in a number of countries, including Canada, have encouraged the development of 'less hazardous' tobacco products.[3] In this context, 'hazard' has and continues to be implicitly equated with 'contents', i.e. cigarettes producing smoke with lowered concentrations of tar and other toxic compounds having a reduced potential hazard in comparison with other brands. Unfortunately, the potential health benefit from switching to low yield cigarettes is not realized by many smokers.[4] Most will obtain the amount of nicotine required for 'satisfaction' independent of the stated yield.[5-7]

There are a number of simple manoeuvres which smokers can use to obtain more nicotine from a cigarette. For example, the nicotine 'content' of an ultra-low tar cigarette (≤ 1 mg) can be increased as much as 19-fold simply by blocking the ventilation holes found on this type of cigarette.[8] Even when the holes are not blocked, 'contents' may be increased to levels several times those published by government agencies by taking larger puffs of longer duration more frequently.[9]

Since some compensation for nicotine has been reported in most brand switching studies,[6,7,10-12] it has been suggested that cigarette smoking could be made less hazardous by reducing tar and other toxins relative to nicotine.[13] In the United Kingdom the average sales-weighted tar/nicotine (T/N) ratio of cigarette brands has declined from 15.5 in 1972 to 11.5 in 1982.[14] The reasons for this decline are unclear since it has not been part of health policy to officially encourage such a change, for good reasons. A policy to reduce tar yields while maintaining nicotine at current levels assumes that nicotine is neither harmful nor gives rise to harmful by-products. In addition, such a policy also assumes that a reduction in the T/N ratio to smoking machines will be reflected in a decrease in the amount of tar per unit nicotine absorbed by smokers. We have investigated this latter aspect by determining how the T/N ratio changes in response to differences in smoking intensity and type of cigarette. Based on our results the cautionary notes which apply to the use of tar and nicotine yields separately[15] apply to the T/N ratio as well.

7.2.   Potential uses for the tar/nicotine ratio

Monitoring the properties of cigarette smoke

It is clear that the average T/N ratio of cigarettes produced in the UK has decreased since 1972. Since major cigarette manufacturers are multinational it is reasonable to ask whether this trend is isolated to the UK or whether the shift to lower T/N ratios has been more widespread. In 1969 the T/N ratio for all 78 Canadian brands whose yield exceeded 10 mg was 14.5[16,17] which decreased by 9 per cent to 13.2 as of October 1986. These Canadian values or the corresponding sales-weighted estimates cannot be compared with the results from the UK for the following reasons.

While it has been agreed internationally that cigarettes be tested by taking a puff of 35 ml over a period of 2 seconds once a minute, there is no such agreement as to when smoking is to stop. For example, in Canada cigarettes are smoked to a length of 30 mm or to within 3 mm of the filter overwrap when the filter-plus-overwrap is less than 27 mm.[17] This choice for a standard butt length has several important consequences. To begin with, in 1986, 76 brands or close to 70 per cent of all Canadian brands could be smoked to a butt length which was shorter than the standard. This percentage has decreased somewhat from 1978, but still represents a clear majority of Canadian cigarettes (see Fig. 7.1). As a result, for regular length cigarettes (72 mm) with short filters as much as 20 per cent of the available tobacco is not smoked when tested by Canadian standards. Consequently, a direct comparison of yields or average yields is not possible.[17] Also, it is not possible to directly compare T/N ratios for brands of Canadian cigarettes with those available in other countries.

In a study of the relationship of T/N ratios as a function of the amount of unsmoked tobacco the following relationship emerged:[18]

$$\text{T/N ratio} = 20.1 - 0.328x + 0.00598x^2$$

where x is the length of the unsmoked tobacco column (mm). This relationship was highly significant, explaining 98 per cent of the variation in T/N ratios for 10 brands of cigarettes whose standard tar yields ranged from 8 to 18 mg. This study was carried out in 1978 and so the consequence of such a relationship can only be discussed in relative terms. For example, when cigarettes in the 1978 study were smoked to within 3 mm of the filter-plus-overwrap which is the effective US standard,[17] the T/N ratio was 19.2. Under Canadian standard conditions this ratio was 17.4 for a decrease of about 9 per cent. Thus, the T/N ratio depends on the amount of tobacco left unsmoked during testing or, in the case of smokers, how much tobacco is discarded. This source of variation in the T/N ratio has serious implications for the prediction of the amount of tar absorbed based on levels of plasma nicotine.[12]

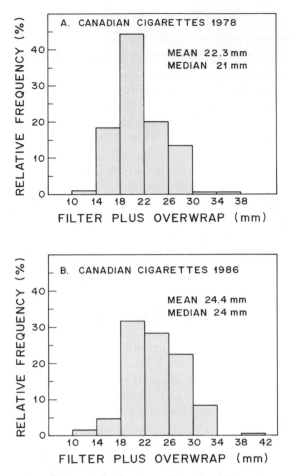

Fig. 7.1. Average, median and distribution of length of filter-plus-overwrap for all
Canadian cigarette brands available in 1978 and 1986.

Predicting tar absorption

In 1976 as part of a butt length study a convenience sample of over 5000
cigarette butts was collected at multiple commercial locations at various
times in the Canadian provinces of Ontario and Newfoundland. Although
this data is not current, it does suggest that there is considerable variability
in the amount of tobacco discarded by smokers (Fig. 7.2). This variability is
not considered when T/N ratios are used to determine an index of tar
absorption (TI). This index is given by:[12]

$$TI(Nic) = \frac{Plasma}{nicotine} \times \frac{Tar\ yield\ of\ cigarette}{Nicotine\ yield\ of\ cigarette}$$

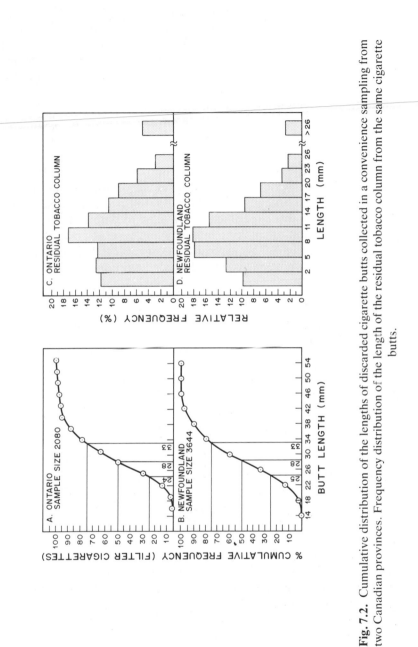

Fig. 7.2. Cumulative distribution of the lengths of discarded cigarette butts collected in a convenience sampling from two Canadian provinces. Frequency distribution of the length of the residual tobacco column from the same cigarette butts.

Assuming an average plasma nicotine of 38.3 ng/ml for middle tar smokers,[12] a range of values for residual column length of from 1 to 26 mm (Fig. 7.2) and the relationship of the T/N ratio to column length noted above, the corresponding values for TI range from a high of 758 ng/ml to a low of 598 ng/ml, a difference of 21 per cent. A decrease in TI of 27.3 per cent has been reported for non-low tar smokers who switched to low tar cigarettes.[12] Even after allowing for variations in butt length, T/N ratios may change considerably depending on how a cigarette is smoked.

In 1986 we reported the results of a study of cigarette yields for 10 brands of Canadian cigarettes smoked under 27 different conditions.[9] These were chosen to represent a wide range of smoking behaviours from 'light' to 'intense', where 'light' means a total smoke volume of about 200–300 ml per cigarette and 'intense' about 900–1000 ml per cigarette. Tar and nicotine yields from this experiment were used to determine how T/N ratios vary as a function of smoking intensity and type of cigarette. The results summarized in Fig. 7.3 are typical for non-vented cigarettes and for ultra-light cigarettes whose tar yield is 2 mg or less. For the 10 mg brand

$$\text{T/N ratio} = 11.5 + 0.00276 \, V$$

where V represents total volume per cigarette (ml) and for the 2 mg brand

$$\text{T/N ratio} = -0.151 + 0.220 \, V + 1.22 \times 10^{-5} V^2.$$

Both regressions are highly significant ($P < 0.005$]. For the ultra-light brand both the linear and quadratic terms are necessary with the quadratic

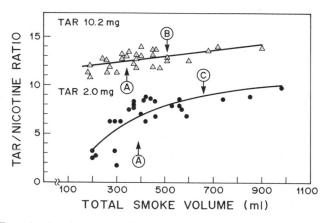

Fig. 7.3. Tar nicotine ratio as a function of the total smoke volume taken for analysis. 'A' represents the volume under standard conditions. 'B' and 'C' the average total volume of inhalation for a middle tar and a low tar cigarette, respectively.[19]

relationship explaining 64 per cent of the variability in T/N ratios. This finding also has implications for potential uses of the T/N ratio.

Multiplication by a constant value for the T/N ratio in the tar index calculation assumes that either individuals obtain the same T/N ratio as the smoking machine or that the smoking behaviour of groups results in an average T/N ratio similar to that found when cigarettes are machine smoked under standard conditions. Neither assumption is likely to be correct. As shown in Table 7.1, allowing for the possibility of differences in smoking intensity the tar index for an average plasma nicotine of 38.3 ng/ml, could be as low as 200 ng/ml for a light smoker of an ultra-light brand or as high as 544 ng/ml for an intense smoker of a middle tar brand.

7.3. Yields as indicators of relative risk to health

Most past studies of smoking behaviour in relation to brand characteristics, including our own,[6,7] have made use of volunteer subjects. For this type of individual, smoking a reduced yield brand appears to result in some decrease in smoke absorption, but not to the extent anticipated based on brand characteristics.[6,7] In order to extend the ability to generalize such findings, we have recently carried out a similar study of smoke absorption, but with a randomly selected non-volunteer sample.

The sampling frame for this experiment was the 1984 tax assessment role for Kitchener, Ontario, Canada (population approximately 145 000). Household addresses were chosen by randomly selecting a page from the assessment role then randomly determining the starting position on that page. A coin toss was used to determine whether to move forward or backward from the starting point until 20 non-commercial addresses had been accumulated. This procedure was repeated 150 times and resulted in the selection of 2891 households.

Table 7.1. Selected values for an index of tar absorption based on variations in the T/N ratio

Smoking intensity*	Tar index (ng/ml)†	
	Middle tar	ultra-low tar
Light (292 ml)	472	200
Standard (350 ml)	478	232
Average‡ (512 ml)	495	303
Intense (970 m)	544	372

*Expressed in terms of total volume of smoke per cigarette.

†The values in the body of this table assume an average of 38.3 ng/ml for plasma nicotine.[12] 'Middle' tar refers to a brand whose nominal delivery exceeds 10 mg. Ultra-low means tar less than or equal to 2 mg.

‡Based on values cited in reference 19, Table 7.2.

Households were visited in the early evening (3–9 p.m.) by trained interviewers. If there were smokers living at that address, one was randomly selected and a request made for either an interview at that time or at some future date. Subjects were asked not to smoke or drink during the interview which lasted at least 15 minutes. At that time a consent form was signed and samples of expired air and saliva obtained.[6] Each subject was paid a total of $5 for participating.

The 2891 addresses surveyed resulted in a sample of 736 participating smokers. Comparisons of sample characteristics such as age, sex, education, levels of cigarette consumption, and type of cigarette smoked with those reported for the province of Ontario[20] did not reveal any major differences.

Numbers for tar and perceived risk to health

Based on questionnaire responses, 47 per cent of 730 subjects stated that they were concerned about the effects of smoking on their health and had tried to lessen the impact in some way (Table 7.2). When probed for specifics, 32 per cent of 441 stated that in order to reduce the impact of smoking on their health they had switched to a lighter/lower/weaker tar brand. It is not too surprising then when subjects were asked to rate the hazards to health of a full flavoured/18 mg brand and a light/3 mg brand, the average score for the 3 mg description was 4.75 while 18 mg was rated as 8.3 (Table 7.3). The extent of the difference between the scores for 3 mg and 18 mg is better determined from the appropriate bar graphs (see Fig. 7.4). The modal value for 3 mg was 3 while the modal value for the 18 mg description was 10. These results indicate that many smokers perceive a quantitative relationship between stated tar yield and risk to health (low tar meaning 'less hazardous'). This finding is one possible explanation for the proliferation of Canadian brands in which the words light and mild

Table 7.2. Tabulation of general health concerns in relation to efforts to reduce smoking related risks*

	Concerned		
Attempted to lessen impact	Yes	No	Totals
Yes	334 (46%)†	72 (10%)	406
No	184 (25%)	140 (19%)	324
Totals	518	212	730

*Response to the questions, 'First. Are you concerned about possible effects of smoking on your health' and 'Have you consciously attempted to lessen the impact of smoking on your health by changing *how* or *what* you smoke in any way'.

†Values are percentages based on 730 responses out of a total sample size of 736.

are part of the brand description (47 per cent of 156 brands). Given the consumers perception of tar numbers, it is unfortunate that in Canada light brands may deliver as little as 4 mg of tar, but as much as 14 mg. Similarly, the range for extra-mild brands is from 0.4 to 12 mg (Table 7.4).

Table 7.3. A summary of the risk to health as perceived by smokers in relation to brand descriptors and tar yields

Grouping	Descriptor*				
	Light	18 mg	Mild	Full flavoured	3 mg
Entire sample	5.38 (690)†	8.30 (605)	5.60 (683)	7.86 (619)	4.75 (632)
Switchers‡					
to low	5.55 (74)	8.86 (72)	5.70 (74)	8.26 (65)	4.92 (74)
to high	5.00 (11)	8.70 (10)	5.64 (11)	8.64 (11)	4.45 (11)
Long-term‡					
low	5.50 (86)	8.47 (83)	5.74 (86)	8.21 (78)	4.86 (84)
middle	4.93 (80)	8.29 (66)	5.44 (82)	7.37 (69)	4.59 (71)
high	5.39 (247)	7.76 (204)	5.54 (240)	7.40 (219)	4.75 (216)

*Responses to the question 'on the same 0–10 scale that we used before where 0 means no health risk and 10 means very serious health risk, what rating would you give a brand described as . . .'.

†Number of responses out of the sample of 736 smokers who participated in this survey.
‡See Table 7.5 for a definition of these terms.

Table 7.4. A comparison of brand descriptors and tar yields of brands of Canadian cigarettes*

Category	Brand designation	Tar yields (mg)			
		No.	Mean	Median	Range
A	Light (s)	28 (18)†	10.7	11	4–14
B	Ultra-light Extra-light	13 (4)	7.6	8	1–11
C	Mild	2 (2)	11.0	11	—
D	Extra-mild Ultra-mild Special mild	30 (23)	6.6	6.5	0.4–12
E	Others	83 (73)	12.8	14	1–18
	All Brands	156 (120)	10.8	11	0.4–18

*List produced by the Canadian Tobacco Manufacturers Council (CMTC) dated June 30, 1986.

†Values are number of brands based on the CMTC list of March 31, 1983.

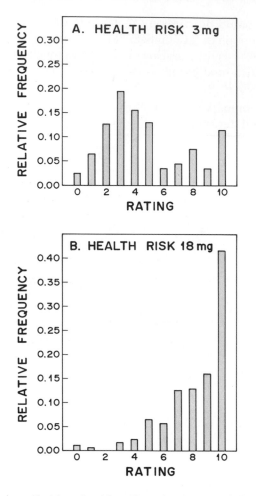

Fig. 7.4. Distribution of risk-to-health ratings for low tar (A) and high tar (B) cigarettes.

Brand characteristics and smoke absorption

To determine if low tar might mean less risk we compared measures of smoke uptake in subjects classified by sex, tar level, and stability of smoking behaviour (brand switchers *v.* non-switchers; see Table 7.5 for a definition of terms). With respect to brand characteristics, the average tar yield for both long-term low tar smokers and those who switched to low tar was 7 mg while the average for long-term high tar smokers was 15 mg. Long-term smokers had smoked their current brand for an average of 6 years (low tar) and 10 years (high tar) while switchers reported smoking current brand for an average of 1.5 years.

Table 7.5. Definition of smoking categories*

Category	Smoke current brand (months)	Previous tar (mg)	Current tar (mg)
1. Switched to low tar brand	≤ 36	≥ 13	≤ 10
2. Switched to high tar brand	≤ 36	≤ 10	≥ 13
3. Long-term low tar smoker	> 36	—	≤ 10
4. Long-term middle tar smoker	> 36	—	11, 12
5. Long-term high tar smoker	> 36	—	≥ 13

*In general, 'switch' means change to a brand with a different name. Switching may take place unknown to the smoker if the manufacturer makes a major change.

Since specific contrasts had been decided upon before the experiment, average consumption adjusted values for indicators of smoke absorption were compared by means of simple t tests. The contrast of females who always smoked low tar cigarettes with those who switched to low tar was highly significant for both breath CO (52 ppm v. 41 ppm; $P < 0.001$) and saliva cotinine 320 ng/ml v. 259 ng/ml; $P < 0.05$). However, females who always smoked low tar cigarettes had higher average results for breath CO than did those who always smoked high tar brands (52 ppm v. 43 ppm; $P < 0.001$). All other contrasts for the data subdivided by sex were not significant.

When the sex of the smoker is ignored (Table 7.6) the comparison of breath CO for long time low tar smokers with those who switched to low was the only contrast which was significant (49 ppm v. 41 ppm; $P < 0.001$).

Table 7.6. Adjusted levels of cotinine in saliva and CO in exhaled air for a random selection of smokers*

	Salivary cotinine (mg/ml)		
	Low tar	High tar	Average
Switcher	273 (65)†	213 (8)	266 (73)
Non-switcher	297 (77)	302 (200)	301 (277)
Average	286 (142)	299 (208)	
	Breath CO (ppm)		
Switcher	41.3 (78)	38.4 (9)	41.0 (87)
Non-switcher	49.3 (80)	45.6 (219)	46.6 (299)
Average	45.3 (158)	45.3 (228)	

*Adjusted for level of consumption.
†Number of subjects.

When both sex and brand loyalty are ignored long-term low tar and long-term high tar smokers had identical average values for breath CO (45.3 ppm). When smokers were categorized based on brand loyalty alone those who had switched down absorbed on the average 12 per cent less nicotine and 12 per cent less carbon monoxide than did those who had smoked the same brand for more than 3 years. If this finding can be generalized, those who switch down, particularly females, may experience a small but tangible health benefit when compared with those who continue to smoke the same brand. Since there were only 11 smokers who reported switching up, this group has not been included in the analysis.

Interpretation of published tar and nicotine yields

There are a number of well known explanations why smokers absorb quantities of smoke which are different from yields under standard conditions. One possibility which generally has not been considered is if smokers perceive that cigarettes contain a specific amount of tar they might also believe that it is not possible to obtain more from a cigarette than the published value. By this reasoning switching to a lower tar brand should result in a health benefit since it should not be possible to inhale more from the new cigarette no matter how it is smoked. Of the 558 individuals who responded, 51 per cent stated that they believed that the numbers which appear on every Canadian cigarette package represents the *maximum* which can be inhaled from that brand of cigarette (Table 7.7). This is obviously incorrect since the maximum based on human smoking behaviour may be as high as 2½ times the published number or even higher.[8,9] The continued publication and advertising of numbers for tar and nicotine reinforces the idea of fixed cigarette contents and perpetuates this fallacy.

7.4. Conclusions

It has been suggested that the hazards associated with smoking could be decreased by maintaining levels of nicotine while reducing the concentration of other hazardous substances such as tar.[13] This approach is strongly supported by evidence that nicotine is the prime reinforcer of the smoking habit.[2] However, it is not clear how reducing the T/N ratio to smoking machines will result in a health benefit to smokers. This investigation has shown that T/N ratios are clearly dependent upon how cigarettes are smoked. As shown in Fig. 7.3, a T/N ratio of 6.1 to a smoking machine may become 9.7 to the smoker who tries to increase the amount of nicotine from an ultra-light cigarette. Figure 7.3 also demonstrates that even under intensive smoking conditions, the T/N ratio for ultra-light cigarette which

Table 7.7. Smoker's perception of tar values as representing the 'maximum' in relation to tar yield of current brand*

Perception of numbers	Tar (mg)		
	≤ 10	11, 12	≥ 13
Represents most	90 (93.8)†	50 (46.1)	143 (143.0)
Can inhale more	95 (91.2)	41 (44.8)	139 (139.0)

*Response to the question 'let's assume for a moment that the tar rating is 10 mg. What does that mean? Is it possible to inhale more than 10 mg tar from each cigarette or does this represent the most you can inhale with this brand?
†Expected values under the hypothesis of independence ($\chi^2 = 0.97$; $P = 0.6165$).

was tested never equalled that of the 10 mg cigarette. For those who will not quit this suggests that there may be some advantage to smoking ultra-light cigarettes provided that the number smoked per day does not increase.

The continued misuse and misinterpretation of cigarette yields indicates that the method of communicating brand characteristics to smokers should be changed. It may be time to acquaint the smoker and researcher alike with the variable nature of yields by publishing a range of values for each brand.[9] A more radical approach would be to stop publishing numbers altogether, replacing them with some other more realistic system. One possibility would be to use colour codes where the colour represents filter tip stain when cigarettes are smoked under a number of conditions.[15] We have investigated this possibility by actually measuring the amount of colour on pads and filters after the deposition of various amounts of tar. The resulting numbers are readily translated by printers so that the colours can be accurately reproduced and used to represent deliveries in terms which smokers may better understand.

Acknowledgements

Financial support was provided by a grant from the National Health Research Development Program (NHRDP), Health and Welfare, Canada and contracts from the Tobacco Products Unit, Health and Welfare, Canada.

References

1. US Surgeon General (1979). *Smoking and health*, DHEW Publication Number (PHS) 79-50066. US Department of Health, Education and Welfare. Washington DC.
2. Henningfield, J. E. (1984). Pharmacologic basis and treatment of cigarette smoking. *Journal of Clinical Psychology*, **45**, 24–34.

3. Independent Scientific Committee on Smoking and Health (1983). Third report of the ISCSH HMSO, London.
4. Rickert, W. S. (1983). Less hazardous cigarettes, fact or fiction? *New York State Journal of Medicine*, **83**, 1269–72.
5. Rickert, W. S. and Robinson, J. C. (1981). Estimating the hazards of less hazardous cigarette. II. Study of cigarette yields of nicotine, carbon monoxide and hydrogen cyanide in relation to levels of cotinine, carboxyhemoglobin and thiocyanate in smokers. *Journal of Toxicology and Environmental Health*, **7**, 391–403.
6. Robinson, J. C., Young, J. C., Rickert, W. S., Fey, G., and Kozlowski, L. T. (1983). A comparative study of the amount of smoke absorbed from low yield ('Less Hazardous') cigarettes part 1: non invasive measures. *British Journal of Addiction*, **77**, 383–97.
7. Robinson, J. C., Young, J. C., and Rickert, W. S. (1985). Maintain levels of nicotine but reduce other smoke constituents. A formula for 'less hazardous' cigarettes. *Preventive Medicine*, **13**, 437–45.
8. Kozlowski, L. T. (1983). Perceiving the risk of low-yield ventilated filter cigarettes: The problem of hole blocking. In *Proceedings of the International Workshop on the Analysis of Actual vs Perceived Risks* (ed. Covello, V., Flamm, W. G., Rodericks, J., and Tardiff, R.). Plenum, New York.
9. Rickert, W. S., Collishaw, N. E., Bray, D. F., and Robinson, J. C. (1986). Estimates of maximum, or average cigarette tar, nicotine and carbon monoxide yields can be obtained from yields under standard conditions. *Preventive Medicine*, **15**, 82–91.
10. US Surgeon General (1982). *The health consequences of smoking. Cancer*. US Department of Health and Human Services, DHHS (PHS) 82-50179.
11. Benowitz, N. L., Jacob, P., Yu, L., Talcott, R., Hall, S., and Jones, R. T. (1986). Reduce tar, nicotine and carbon monoxide exposure while smoking ultra low but not low yield cigarettes. *Journal of the American Medical Association*, **256**, 241–6.
12. Russell, M. A. H., Jarvis, M. J., Feyerabend, C., and Saloojee, Y. (1986). Reduction of tar, nicotine and carbon monoxide intake in low tar smokers. *Journal of Epidemiology and Community Health*, **40**, 80–5.
13. Russell, M. A. H. (1976). Low tar medium nicotine cigarette: A new approach to safer smoking. *British Medical Journal*, **1**, 1430–3.
14. Jarvis, M. J. and Russell, M. A. H. (1985). Tar and nicotine yields of U.K. cigarettes 1972–1983: Sales weighted estimates from non industry sources. *British Journal of Addiction*, **80**, 429–34.
15. Fourth Scarborough Conference on Preventive Medicine (1985). Is there a future for lower tar yield cigarettes. *Lancet*, **ii**, 1111–4.
16. Forbes, W. F., Robinson, J. C., and Stanton, M. (1969). Tar and nicotine retrieval from cigarettes available in Canada. *Cancer*, **23**, 910–2.
17. Rickert, W. S., Robinson, J. C., and Young, J. C. (1980). Estimating the hazards of less hazardous cigarettes. I. Tar, nicotine, carbon monoxide, acrolein, hydrogen cyanide and total aldehyde deliveries of Canadian cigarettes. *Journal of Toxicology and Environmental Health*, **6**, 351–65.
18. Young, J. C., Rickert, W. S., and Robinson, J. C. (1981). A study of chemical deliveries as a function of cigarette butt length. *Beitrage zur Tabakforschung*, **11**, 87–95.

19. Nil, R., Buzzi, R., and Battig, K. (1986). Effects of different cigarette smoke yields on puffing and inhalation: Is the measurement of inhalation volumes relevant for smoke absorption. *Pharmacology Biochemistry and Behaviour*, **24**, 587–95.
20. Joseda, D. (1983). *Smoking behaviour of Canadians*. Health Promotion Directorate, Health and Welfare Canada, Cat. No. H39-66/1985E ISBN 0-662-13947X.

8

Estimating the extent of compensatory smoking

ALISON STEPHEN, CHRISTOPHER FROST, SIMON
THOMPSON, and NICHOLAS WALD

Abstract

*When smokers switch to a cigarette brand having a lower tar yield than
their usual brand they typically 'compensate' for the reduction in yield
by increasing the extent to which they inhale. This paper defines
compensation and estimates it in 17 published studies that used carbon
monoxide and nicotine as tobacco smoke markers. The average value
of compensation based on carbon monoxide was 73 per cent and that
based on nicotine intake was 66 per cent. A method of estimating tar
intake from the intake of other smoke constituents is suggested and
compensation based on tar (CO derived) estimated to be 60 per cent.
Using this figure we calculate that smokers who reduce the tar yield of
their cigarettes by half will, on average, reduce their intake of tar by 24
per cent.*

8.1. Introduction

It is recognized that smokers who switch to cigarettes with a lower tar yield
inhale more smoke to compensate for the reduction in yield. The International Agency for Research on Cancer in their review on tobacco
smoking[1] summarize a number of studies that have investigated this effect
and conclude that all the studies are consistent in demonstrating some
degree of compensation when switching to low tar cigarettes. Conversely,
those smokers switching to higher yield cigarettes reduce their extent of
inhaling.

The IARC report did not define 'compensation' or produce an overall
summary estimate of its magnitude. Here, we propose a definition of
compensation and combine data from published studies to estimate its
value.

8.2. Methods

Compensatory smoking

A smoker's intake of tar, nicotine, carbon monoxide, and other smoke components need not change directly in proportion to the change in yield of the various smoke components as predicted by a smoking machine when changing from one brand of cigarette to another. If the change is to a low yield cigarette the smoker is likely to inhale more smoke to compensate for the lower concentration of one or more of the smoke constituents.

The intake of smoke constituents can be measured by markers of smoking. In our analysis we restrict attention to the three most commonly used markers. Carboxyhaemoglobin[2-10,18] (COHb) and breath carbon monoxide[12-16] (BCO) are both markers of carbon monoxide intake and plasma nicotine[3-6,8,11,13,16-18] is a marker of nicotine intake. Cotinine, a metabolite of nicotine, is now recognized as a better marker of nicotine intake than plasma nicotine,[19] but few studies have measured cotinine. There is no adequate marker of tar; tar intake can only be estimated by inference from the intake of markers of other smoke components.

Both COHb and BCO are present in non-smokers, average levels being, respectively, 0.7 per cent and 6 ppm.[20,21] These background levels of COHb and BCO must be subtracted from COHb and BCO levels to provide an estimate of intake attributable to active smoking. In the rest of this paper COHb and BCO relate to their respective levels above background. Background levels of nicotine are negligible and so no adjustment of plasma nicotine is necessary.

Definition of compensation

Definitions of zero (0 per cent) and full (100 per cent) compensation are straightforward. Consider, for example, the changes in COHb that result from changes in CO yield as a smoker changes cigarette brand while smoking the same number of cigarettes at the same intervals. Zero compensation occurs if the smoker inhales to the same extent before and after the change of brand. If the CO yield of the new cigarette is half that of the old one, then the COHb level will also be halved. Full compensation occurs when the smoker changes his or her degree of inhalation to maintain a constant COHb concentration.

Consider a 'partial compensator' who reduces the CO yield of his or her cigarette by 50 per cent, but whose COHb level falls by only 20 per cent, to 80 per cent of its original level. At what rate is this smoker compensating? To answer this question it is useful to consider what might happen if the smoker were to further reduce the CO yield of his or her cigarette to 50 per cent of this new value (i.e. to 25 per cent of the original CO yield). If the smoker continues to compensate to the same degree, the COHb level would be reduced by a further 20 per cent to 64 per cent (0.64 = 0.8 × 0.8) of its original level (Table 8.1). A further reduction of 50 per cent in CO yield would be expected to lead to a further reduction of 20 per cent in COHb level to 0.512 (0.64 × 0.8) of its original value. Since a given proportional change in yield (50 per cent) is always reflected in a given proportional change in marker (20 per cent), the relationship is linear when both marker and yield are expressed on logarithmic scales. (Fig. 8.1

Table 8.1. Hypothetical smoker compensating for a reduction in CO yield at a constant rate

	CO yield (expressed as a proportion of initial yield)	COHb (expressed as a proportion of initial yield)
Initial levels	1	1
After first reduction in yield	0.5	0.8
After second reduction in yield	0.25 (0.5 × 0.5)	0.64 (0.8 × 0.8)
After third reduction in yield	0.125 (0.25 × 0.5)	0.512 (0.64 × 0.8)

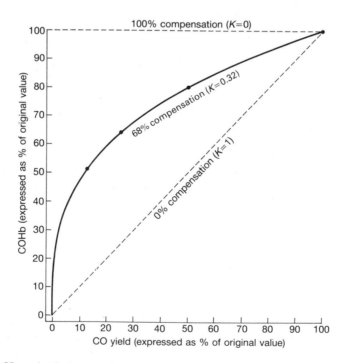

Fig. 8.1. Hypothetical example of a smoker compensating at a constant rate illustrated on an arithmetic scale.

illustrates this relationship on an arithmetic scale, Fig. 8.2 on a logarithmic scale.) The relation is defined by the following equation when both marker and yield are expressed as proportions of their initial values:

$$\log(\text{marker}) = K \times \log(\text{yield}) \tag{1}$$

where $K = \log(0.8)/\log(0.5) = 0.32$ in this example.
On arithmetic scales this relationship is:

$$\text{marker} = (\text{yield})^K \tag{2}$$

For any smoker compensating uniformly with changes in cigarette brand and not changing the number of cigarettes smoked or the intervals between them, marker and yield will be related as in equations (1) and (2) with K reflecting compensation. K is defined as follows for any change in yield:

$$K = \frac{\log(\text{marker; expressed as a proportion of initial value})}{\log(\text{yield; expressed as a proportion of initial value})} \tag{3}$$

K is 1 for 'zero compensation' (because numerator and denominator will be equal) and 0 for 'full compensation' (because numerator will be zero) and lie between 0 and 1, for 'partial compensation'. We can therefore define compensation to be $1 - K$ with K as in eqn (3) above and rewrite equation (1) as:

$$\log(\text{marker}) = (1 - C) \times \log(\text{yield}) \tag{4}$$

where C is compensation.
For the hypothetical smoker discussed above compensation is 68 per cent $[0.68 = 1 - \log(0.8)/\log(0.5)]$. As a second example 50 per cent compensation arises if a 75 per cent reduction in yield (i.e. to 25 per cent of initial value) is associated with a reduction of 50 per cent in intake.

Extent of inhaling

The compensation eqn (4) can be rearranged to:

$$\log(\text{marker}/\text{yield}) = -C \times \log(\text{yield}) \tag{5}$$

This ratio of marker to yield is a measure of the extent of inhaling reflecting the aggregate of the three main components of inhaling: number of puffs, puff volume, and depth to which the smoke is inhaled into the lungs. For example, if a smoker switches to a cigarette having half the CO

Alison Stephen et al.

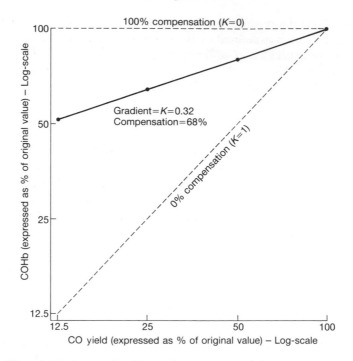

Fig. 8.2. Hypothetical example of a smoker compensating at a constant rate illustrated on a logarithmic scale.

yield of his original brand, but maintains a constant intake of CO (100 per cent compensation) the smoker will be smoking the new cigarettes twice as hard (for CO) as the old ones. Equation (5) shows that the 'Extent of inhaling' ratio is linearly related to yield via (−) compensation when both are expressed on logarithmic scales. If compensation is zero the ratio is constant with respect to changes in yield, while if compensation is 1 the ratio changes in inverse proportion to changes in yield so that, as above, a 50 per cent reduction in yield would be associated with a two-fold increase in the extent of inhaling.

Summarizing compensation over many brand changes

When a smoker changes cigarette brand the extent of compensation can be estimated using eqn (3). In many studies of compensatory smoking more than one change of brand is involved. The compensation rate for each of these brand changes is unlikely to be exactly the same and it is often useful to calculate a compensation rate that summarizes all the changes made by that smoker. Simple regression can be used to calculate such a summary figure because eqn (4) which defines compensation is a special case of the more general form:

$$\log(\text{marker}) = a + (1 - C) \times \log(\text{yield}) \tag{6}$$

where C is compensation as before, and a is a scaling constant.

This equation is correct whatever units are used to measure the marker and yield [a is a constant whose value is zero when marker and yield are expressed as proportions of some initial value: i.e. eqn (4)]. From equation (6) a summary compensation value can be calculated for individuals who change brand more than once by regressing log(marker) on log(yield) to find the best straight line through the points. The gradient of this line will be an estimate of one minus compensation.

Compensation for tar, nicotine, and CO

In calculating compensation for particular smoke constituents no statements about which constituents determine compensation can be made. The smoker who switches to a low tar cigarette may be inhaling more deeply because of the reduced CO, tar, or nicotine yield, or indeed, because some other constituent of smoke is reduced.

Compensation will be different for different smoke constituents. After a change of cigarette brand the proportional change in the extent of inhaling is the same for each constituent. Compensation for the various constituents will vary because the proportional changes in their respective yields will be different.

Although tar intake cannot be measured directly, inferences about compensation for tar can be made from the intake of another smoke component such as CO. In the above example (CO yield reduced by 50 per cent and COHb reduced by 20 per cent), the smoker inhales more smoke after the change of brand than before. It is therefore likely that the smoker's tar intake (per unit of tar yield) will also be increased. The most reasonable assumption seems to be that proportionate changes in the extent of inhaling CO (COHb/CO yield) are mirrored in equal proportionate changes in the extent of inhaling tar (tar intake/tar yield). Equation (5) shows that proportional change in the extent to which a particular smoke component is inhaled is linearly related to proportional change in the yield of that smoke component, their ratio being a measure of compensation for the change in yield of that smoke component. It follows that, with the above

assumption, compensation for tar (CO based) can be estimated from the
following equation:

$$C_{tar} = C_{CO} \times \frac{\log(\text{CO yield; expressed as a proportion of initial value})}{\log(\text{tar yield; expressed as a proportion of initial value})} \quad (7)$$

where C_{tar} is the compensation for tar (estimated from COHb), and C_{CO} is
the compensation for CO.

Compensation in individual studies

Two types of study have measured compensation: intervention studies and
observational studies. Compensation, as defined above, can be calculated
in each study. In our analysis we restricted attention to studies measuring
either (1) COHb or breath CO marker levels and CO yield levels, or (2)
plasma nicotine and nicotine yield levels. The studies are referenced in
Tables 8.2[2–10,12–16,18] (markers of CO) and 8.3[3–6,8,11,13,16–18] (nicotine).

Intervention studies having a cross-over design. In an intervention study
the participants smoke their usual cigarettes for a period of time and then

Table 8.2. Studies of CO compensation

Study	Marker	Number of subjects	Compensation (%)	
			Unadjusted	Adjusted for cigarette consumption
Intervention studies				
Turner et al. (1974)[2]	COHb	10	38.6	32.9
Ashton et al. (1979)[3]	COHb	12	91.7	86.8
Hill et al. (1980)[4]	COHb	3	106.8	90.8
Stepney (1980)[12]	BCO	19	42.5	40.5
Fagerstrom (1982)[13]	BCO	12	62.6	63.6
Robinson et al. (1982)[14]	BCO	16	94.3	92.7
Russell et al. (1982)[5]	COHb	12	84.8	66.6
Benowitz et al. (1984)[6]	COHb	11	93.7	78.7
Robinson et al. (1984)[7]	COHb	16	68.4	63.9
Benowitz et al. (1986)[8]	COHb	22	81.3	76.1
Observational studies				
Rickert et al. (1981)[9]	COHb	31	73.7	87.3
Stepney (1982)[15]	BCO	78	59.0	50.5
Ebert et al. (1983)[16]	BCO	76	96.1	91.2
Wald et al. (1984)[10]	COHb	2455	100.1	100.1
Russell et al. (1986)[18]	COHb	392	72.8	80.5

Table 8.3. Studies of nicotine compensation

Study	Number of subjects	Compensation (%)	
		Unadjusted	Adjusted for cigarette consumption
Intervention studies			
Russell *et al.* (1975)[17]	10	57.2	47.6
Ashton *et al.* (1979)[3]	12	76.5	72.6
Hill *et al.* (1980)[4]	4	76.2	51.0
Fagerstrom (1982)[13]	12	107.3	101.8
Russell *et al.* (1982)[5]	12	32.1	25.4
Benowitz *et al.* (1984)[6]	11	84.2	68.0
Benowitz *et al.* (1986)[8]	22	69.9	66.1
Observational studies			
Russell *et al.* (1980)[11]	222	76.8	88.0
Ebert *et al.* (1983)[16]	76	78.0	72.6
Russell *et al.* (1986)[18]	83	62.7	68.2

change to a brand with different tar, nicotine, or CO yields for a similar length of time. During both periods levels of the markers – COHb, breath CO, and/or nicotine – are measured. Compensation can be calculated directly since such studies compare the same individuals smoking different cigarette brands. All of the studies are small, none having more than 20 participants and all are similar in design, varying primarily in their duration of intervention, the longest being 10 weeks and the shortest 1 day (see Tables 8.2 and 8.3).

As defined above, compensation relates to an individual. In most studies, data on individuals were not available and so individual estimates of compensation could not be derived. However, the same concept of compensation can be applied to the average levels of marker and yield for the group of smokers before and after switching brands to produce an average estimate of compensation.

In some of the studies only one reduction in yield was involved. Calculation of compensation was then straightforward from eqn (3). In other studies measurement was made of the marker concerned in relation to more than one change of yield. Average levels of the marker were presented for individuals smoking cigarettes with three or more different yields. In these studies, a summary statistic was calculated by regressing (unweighted) the average marker level on the average yields using logarithmic scales [eqn (6)]. Compensation is estimated to be 1 minus the gradient of the regression line. The approach is illustrated using data from the study by Stepney in Fig. 8.3 which involved three brand comparisons.

Alison Stephen et al.

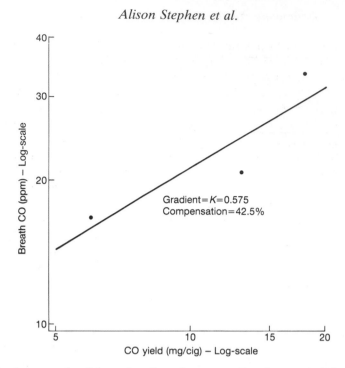

Fig. 8.3. An example of the estimation of compensation from a study by Stepney. (Breath CO concentrations associated with smoking three brands of cigarette.)

Estimates of compensation for all the studies we have included in our analysis are given in Tables 8.2 and 8.3.

Observational studies. In observational studies the intake of smoke constituents in habitual smokers of low yield cigarettes is compared to the intake in smokers of higher yield cigarettes. In such circumstances compensation can be estimated in the same way as in the intervention studies, but comparison is made between separate groups of individuals smoking different cigarette brands rather than between the same individuals smoking different cigarette brands. Random error is, in general, smaller than in the intervention studies because larger groups of subjects are involved in the observational studies.

Compensation was calculated as for the intervention studies using average levels of marker and yield in the different yield groups. Estimates of compensation for each study are displayed in Tables 8.2 and 8.3.

Adjustment for cigarette consumption

The level of a marker of smoke intake is related not only to the brand of cigarette smoked, but also to the number of cigarettes smoked, and estimates of compensation therefore need to be adjusted for cigarette consumption. All but one of the studies recorded cigarette consumption (either by number of cigarettes smoked on the day that intake was measured or by average consumption) at the various yield levels. The study by Wald *et al.*[10] adjusted marker levels for cigarette consumption and so no further adjustment was necessary in that study. Typically, cigarette consumption was greatest in the lowest yield groups as smokers tried to compensate by smoking more cigarettes as well as inhaling more smoke, but this difference in cigarette consumption was usually small. Nevertheless, if changes in cigarette consumption are not taken into account compensation will tend to be slightly overestimated in the studies.

Compensation estimates were adjusted for cigarette consumption by fitting the following unweighted multiple regression model to (1) all the studies of CO intake and (2) all the studies of nicotine intake:

$$\log(\text{marker}) = a + (1 - C) \times \log(\text{yield})$$
$$+ b \times \log (\text{cigarette consumption}) \quad (8)$$

This model extends that used to estimate compensation in the individual studies by adding a cigarette consumption term to eqn (6). *a* and *C* were allowed to vary from study to study, but *b* was held constant across studies, thereby reflecting the 'average' effect of proportional changes in cigarette consumption on intake. The values of *C* estimated by the model are estimates of compensation adjusted for cigarette consumption (Tables 8.2 and 8.3).

8.3. Results

Markers of CO intake – COHb and breath CO (Table 8.2)

Figure 8.4 illustrates the range of values of CO-based compensation (adjusted for cigarette consumption) in the studies analysed. A value of more than 100 per cent indicates that, on average, the marker level rose with decreased yield. The mean value is 73 per cent (standard error = 5 per cent). An unweighted mean is presented to reflect the fact that there is likely to be considerable variation between studies that is not explained by between-subjects random variation. There is considerable variation in the individual study compensation levels, this variation being, as expected, greater in the small scale intervention studies than the large observational studies.

Unweighted multiple regression analysis showed no significant difference in compensation levels between those studies using COHb and those using breath CO as the markers of CO intake, or between the intervention

Alison Stephen et al.

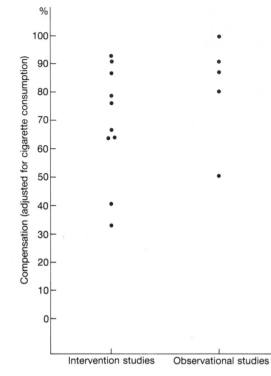

Fig. 8.4. CO-based compensation estimated in 15 studies.

and observational studies. Neither was the duration of the intervention studies significantly related to the estimate of compensation in those studies. In addition, there was no evidence for a systematic relationship between the estimates of compensation and the absolute yields of the cigarettes used in the studies suggesting that compensation is not materially different at different yield levels.

Marker of nicotine intake − plasma nicotine (Table 8.3)

Figure 8.5 illustrates the range of values of nicotine-based compensation (adjusted for cigarette consumption) in the studies analysed. The mean value is 66 per cent (standard error = 7 per cent). Again, neither the type of study, yield levels in the study, nor the duration of the intervention studies was significantly related to the estimate of compensation.

Compensation for tar

Across all comparison groups in all studies average tar, nicotine and CO yield levels were all highly correlated (correlation coefficients of 0.93

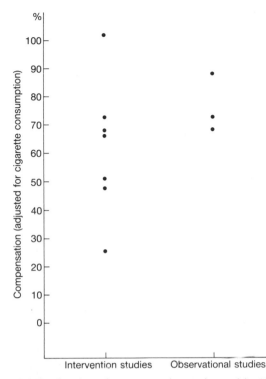

Fig. 8.5 Nicotine-based compensation estimated in 10 studies.

between tar and nicotine, 0.94 between CO and tar, 0.80 between CO and nicotine) on logarithmic scales (Fig. 8.6). Because these correlations are so high across studies changes in tar yield must be closely related to changes in CO yield and so an overall average value for tar compensation can be estimated from the average value for CO compensation and the typical ratio of the proportional change in CO yield to the proportional change in tar yield (eqn (7)).

Regression of tar yields on CO yields (logarithmic scales) for the study groups analysed can provide an estimate of reduction in tar yield associated with a 50 per cent reduction in CO yield. However, when two variables vary together, regression techniques tend to underestimate the typical variation of one variable with the other as both variables are subject to random error. Principal components analysis[22] is an alternative statistical technique which avoids the problem – the first principal component of the tar, nicotine, and CO yields (on a logarithmic scale) providing an estimate of the typical variation in all three yields together.* The first principal

*Statistical note: using logarithmic scales, thereby relating proportional changes in variables means that the usual problem of 'scaling' in principal components analysis is avoided.

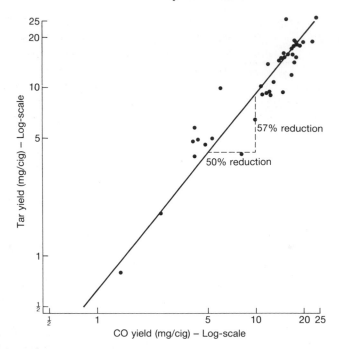

Fig. 8.6 Correlation between average tar and CO yields in the different comparison groups of the studies analysed. First principal component represents the typical change in tar yield associated with a change in CO yield.

component of the nicotine, tar, and CO yields in the comparison groups suggests that a 50 per cent drop in CO yield is associated, typically, with a 49 per cent drop in nicotine yield and a 57 per cent drop in tar yield. Compensation for tar can be estimated as 60 per cent $[0.73 \times \log(0.5)/\log(0.43)$, eqn (7)] from CO-based compensation or as 53 per cent $[0.66 \times \log(0.51)/\log(0.43)]$ from nicotine-based compensation.

8.4. Discussion

All the studies on compensatory smoking illustrate that compensation occurs. Smokers of low tar cigarettes smoke those cigarettes 'harder' and so inhale a greater proportion of smoke constituents than a simple comparison of yield levels would suggest. The studies are also consistent in demonstrating that compensation is not complete. Low tar cigarette smokers inhale less CO and nicotine than high tar cigarette smokers. In this paper we have taken the analysis of compensation further by (1) providing a plausible quantitative definition of compensation and (2) obtaining an overall estimate of compensation for the studies cited using the three most commonly used biological markers of intake, namely COHb, breath CO, and blood nicotine.

We have demonstrated that constant compensation is reflected in a non-linear relationship between marker and yield levels showing that compensation cannot be defined by linearly relating arithmetic changes in yield to arithmetic changes in marker level. Constant compensation is, instead, reflected in proportionate changes in marker with proportionate changes in yield. Thus, when marker and yield are both logarithmically transformed the relationship between the two is linear and compensation can be defined as:

$$1 - \frac{\log\ (\text{marker; expressed as a proportion of level at initial yield})}{\log\ (\text{yield; expressed as a proportion of initial yield})}$$

The average value for CO-based compensation in the studies analysed is 73 per cent, whilst the average value for nicotine-based compensation is 66 per cent.

Under the assumption that CO intake (per unit of CO yield) reflects tar intake (per unit of tar yield) we estimate compensation for tar to be 60 per cent, a figure which is lower than that for CO compensation because switching to low tar cigarettes involves a larger proportionate change in tar yield than in CO yield. A similar assumption that nicotine intake reflects tar intake leads to an estimate of compensation for tar of 53 per cent.

These results are of importance in assessing the health implications of switching to low tar cigarettes. Assuming that the risk of disease is directly related to the estimated change in intake of the smoke component thought to cause that disease we can say that, from our data, a reduction in tar yield of 50 per cent will result in a 24 per cent $[1 - 0.5^{(1-0.60)}$, eqn (2)] reduction in tar intake and the same reduction in the risk of, say lung cancer. This is based on the CO-derived estimate of compensation for tar of 60 per cent. Such an estimate, however, might have to be adapted in the light of other relevant information. For example, the risk of lung cancer might be reduced by more than 24 per cent because when the yield of a cigarette is reduced and the depth of inhaling increased as a result, the tar droplets may be deposited more in the peripheral parts of the lung and less on the proximal bronchial airways.[23] This would lead to a greater reduction in lung cancer, which usually occurs on the proximal bronchial epithelium, than would be expected if the distribution of droplet deposition in the lung did not change with changes in the extent of inhaling. In whatever way our estimates may need to be refined before drawing inferences about the risk of smoking-related diseases, it is reassuring having quantified the relationship between yield and intake of smoke constituents we can conclude that on average switching from cigarettes with relatively high yields to cigarettes with lower yields will lead to a reduced intake of the smoke constituents we have considered even though the reduction in intake is significantly less than is predicted by a smoking machine.

References

1. IARC (1986). *Monographs on the evaluation of the carcinogenic risk of chemicals to humans*, Vol. 38, *Tobacco smoking*. International Agency for Research on Cancer, Lyon.
2. Turner, J. A. McM., Sillett, R. W., and Ball, K. P. (1974). Some effects of changing to low-tar and low-nicotine cigarettes. *Lancet*, ii, 737–9.
3. Ashton, H., Stepney, R. and Thompson, J. W. (1979). Self-titration by cigarette smokers. *British Medical Journal*, **2**, 357–60.
4. Hill, P. and Marquadt, H. (1980). Plasma and urine changes after smoking different brands of cigarettes. *Clinical Pharmacology and Therapeutics*, **27**, 652–8.
5. Russell, M. A. H., Sutton, S. R., Iyer, R., Feyerabend, C., and Vessey, C. J. (1982). Long term switching to low-tar low-nicotine cigarettes. *British Journal of Addiction*, **77**, 145–58.
6. Benowitz, N. L. and Jacob, P. (1984). Nicotine and carbon monoxide intake from high- and low-yield cigarettes. *Clinical Pharmacology and Therapeutics*, **36**, 265–70.
7. Robinson, J. C., Young, J. C., and Rickert, W. S. (1984). Maintain levels of nicotine but reduce other smoke constituents: a formula for 'less-hazardous' cigarettes? *Preventive Medicine*, **13**, 437–45.
8. Benowitz, N. L., Jacob, P., Yu, L., Talcott, R., Hall, S., and Jones, R. T. (1986). Reduced tar, nicotine and carbon monoxide exposure while smoking ultralow, but not low-yield cigarettes. *Journal of the American Medical Association*, **256**, 241–6.
9. Rickert, W. S. and Robinson, J. C. (1981). Estimating the hazards of less hazardous cigarettes. II. Study of cigarette yields of nicotine, carbon monoxide, and hydrogen cyanide in relation to levels of cotinine, carboxyhemoglobin and thiocyanate in smokers. *Journal of Toxicology and Environmental Health*, **7**, 391–403.
10. Wald, N. J., Boreham, J., and Bailey, A. (1984). Relative intakes of tar, nicotine and carbon monoxide from cigarettes of different yields. *Thorax*, **39**, 361–4.
11. Russell, M. A. H., Jarvis, M., Iyer, R., and Feyerabend, C. (1980). Relation of nicotine yield of cigarettes to blood nicotine concentrations in smokers. *British Medical Journal*, **1**, 972–6.
12. Stepney, R. (1981). Would a medium-nicotine low-tar cigarette be less hazardous to health? *British Medical Journal*, **283**, 1292–6.
13. Fagerstrom, K. O. (1982). Effects of a nicotine-enriched cigarette on nicotine titration, daily cigarette consumption, and levels of carbon monoxide, cotinine and nicotine. *Psychopharmacology*, **77**, 164–7.
14. Robinson, J. C., Young, J. C., and Rickert, J. S. (1982). A comparative study of the amount of smoke absorbed from low yield ('less hazardous') cigarettes. Part I: Non-invasive measures. *British Journal of Addiction*, **77**, 383–97.
15. Stepney, R. (1982). Exposure to carbon monoxide in smokers of middle- and low-tar cigarettes. *British Journal of Diseases of the Chest*, **76**, 390–6.
16. Ebert, R. V., McNabb, M. E., McCusker, K. T., and Snow, S. L. (1983). Amount of nicotine and carbon monoxide inhaled by smokers of low-tar, low-nicotine cigarettes. *Journal of the American Medical Association*, **250**, 2840–2.

17. Russell, M. A. H., Wilson, C., Patel, U. A., Feyerabend, C., and Cole, P. V. (1975). Plasma nicotine levels after smoking cigarettes with high, medium and low nicotine yields. *British Medical Journal*, **2**, 414–6.
18. Russell, M. A. H., Jarvis, M. J., Feyerabend, C., and Saloojee, Y. (1986). Reduction of tar, nicotine and carbon monoxide intake in low tar smokers. *Journal of Epidemiology and Community Health*, **40**, 80–5.
19. Hill, P., Haley, N. J., and Wynder, E. L. (1983). Cigarette smoking: Carboxyhemoglobin, plasma nicotine, cotinine and thiocyanate vs. self-reported smoking data and cardiovascular disease. *Journal of Chronic Diseases*, **36**, 439–49.
20. Wald, N. J., Idle, M., Boreham, J., and Bailey, A. (1981). Carbon monoxide in breath in relation to smoking and carboxyhaemoglobin levels. *Thorax*, **36**, 366–9.
21. Wald, N., Idle, M., and Bailey, A. (1978). Carboxyhaemoglobin levels and inhaling habits in cigarette smokers. *Thorax*, **33**, 201–6.
22. Chatfield, C. and Collins, A. J. (1980). *Introduction to multivariate analysis*. Chapman and Hall, London.
23. Wald, N., Idle, M., Boreham, J., and Bailey, A. (1983). Inhaling and lung cancer: an anomaly explained. *British Medical Journal*, **287**, 1273–5.

9

Consistency of nicotine intake in smokers of cigarettes with varying nicotine yields

F. ADLKOFER, G. SCHERER, A. BIBER, W.-D. HELLER,
P. N. LEE, and H. SCHIEVELBEIN

Abstract

In two field studies the within-person variation of nicotine intake was investigated in 303 smokers. In the first study, the serum cotinine level of 200 male smokers was determined in three blood samples from each subject drawn at weekly intervals. In the second study, six blood samples each were taken from 51 male and 52 female smokers at intervals of 4–6 weeks for serum cotinine determination. Both studies revealed that nearly one-third of the smokers take in relatively low amounts of nicotine, which suggests that these subjects do not smoke primarily for nicotine. The nicotine intake varied widely between the smokers, but was rather constant for each smoker. A mathematical compensation model applied to the data of the second study reveals that the amount of nicotine a smoker takes in is determined only to a small extent by the nicotine yield of the brand smoked. From our findings we conclude that a major proportion of the smoking population compensates to a considerable extent, probably for nicotine, and thus might benefit from a decrease in the tar/nicotine ratio in cigarettes.

9.1. Introduction

Many experimental and observational studies have provided evidence suggesting that smokers adjust their inhalation pattern in order to compensate for any changes in cigarette yields as measured by smoking machine.[1] Nicotine is thought to determine the extent of smoke intake.[2] If this is the case, the serum level of its main metabolite, cotinine, should be relatively constant over time within one individual. Apart from a few casual observations[3–5] this question has so far not been investigated systematically in longitudinal studies. In this paper we report on two field studies dealing with nicotine intake over time.

9.2. Subjects and methods

The cigarette smokers were recruited by newspaper advertisements, by posters in shops, factories, public buildings and universities, as well as by word of mouth. Each subject completed a detailed questionnaire on smoking behaviour, life style, occupation and psychosocial characteristics.

Study 1

Two-hundred male cigarette smokers (mean age 26.0 years, range 18–40 years) provided three blood samples each, which were drawn at 1-week intervals on Tuesdays, Wednesdays, or Thursdays to avoid interference with weekend activities. The samples were taken between 4 p.m. and 6 p.m. The subjects were asked about their cigarette consumption on that particular day only during their first visit. They were not informed about the aim of the study which ran under the title 'Lipid Research Study'.

Study 2

Fifty-one male and 52 female cigarette smokers (mean age 26.4 years, range 20–40 years) provided six blood samples each at intervals of 4–6 weeks. On each visit to the laboratory subjects were asked about their cigarette consumption as well as about their food, alcohol, and drug intake over the last 48 hours. In all other respects Study 2 was similar to Study 1.

Analytical methods

Nicotine and cotinine in serum were determined by gas chromatography (GC)[6] in Study 1 and by radioimmuno-assay (RIA)[7] in Study 2. Both methods are similar in precision and accuracy, although RIA values are somewhat higher than those obtained by GC.[8]

Statistical methods

Statistical analysis was carried out using the SAS (Statistical Analysis System) package. Multiple regression techniques were used in Study 2 in order to quantify more precisely the relationship of serum cotinine levels to cigarette consumption and to nicotine delivery. These techniques were based on the model $C = \mu\, N^\alpha\, Y^\beta$ where C, serum cotinine, is related to the number of cigarettes smoked (N) by the factor N^α and to the nicotine yield (Y) by the factor Y^β, α and β may be termed 'compensation indices'. A value of α or $\beta = 0$ implies complete compensation, i.e. no relationship of cotinine to number of cigarettes smoked or to nicotine yield, whereas a value of 1 implies direct proportionality – no compensation. Taking log-arithms this model becomes linear

$$\log C = \log \mu + \alpha \log N + \beta \log Y$$

and permits the use of standard procedures to estimate α, β, and $\log \mu$ directly. Formally, if i represents person and j represents point of time, the following models were tried initially

1. $\log C_{ij} = \log \mu_i$
2. $\log C_{ij} = \log \mu_i + \alpha \log N_{ij}$
3. $\log C_{ij} = \log \mu_i + \alpha \log N_{ij} + \beta \log Y_{ij}$

in all of which only μ_i was assumed to vary from person to person. Effectively, parallel lines were fitted, with the slopes (compensation indices) assumed the same for each person. Further models were tested in which the slopes were allowed to vary for subjects in different normal consumption or normal yield groups or for each subject individually; 0.5 was added to cotinine values before taking logarithms since there were occasional zero values.

All of these analyses in Study 2 were restricted to those subjects for whom normal consumption and yield, as determined from the initial questionnaire, were known, and to those occasions where C, N, and Y were not missing. This left a total of 463 observations on 86 subjects.

In Study 1, 42 subjects were excluded from the analysis because they used other tobacco products in addition to manufactured cigarettes or smoked hand-rolled cigarettes.

9.3. Results

Study 1

Figure 9.1 shows the frequency distribution of serum cotinine values for the three points of time. There is a high inter-individual variation. Interestingly, a considerable number of smokers, about 30 per cent, had low values of less than 100 ng/ml. As shown in Fig. 9.2 A–C, there is a statistically significant correlation between serum cotinine and number of cigarettes smoked per day, although it is relatively weak ($r = 0.41$–0.44). The figures also demonstrate the individual consistency of serum cotinine over time by assigning the smokers to three groups (indicated by the symbols ●, ■, and ▲) according to the cotinine level determined at the first investigation (Fig. 9.2A), using arbitrary borderlines of 70 ng/ml and 200 ng/ml. Figure 9.2B, C shows that most smokers (about 75 per cent) stay within their borderlines,

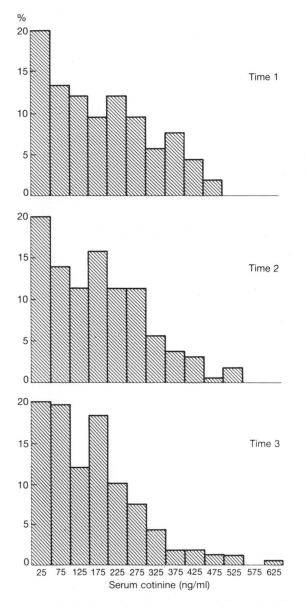

Fig. 9.1. Frequency distribution of cotinine values (Study 1, 158 male cigarette smokers).

F. Adlkofer et al.

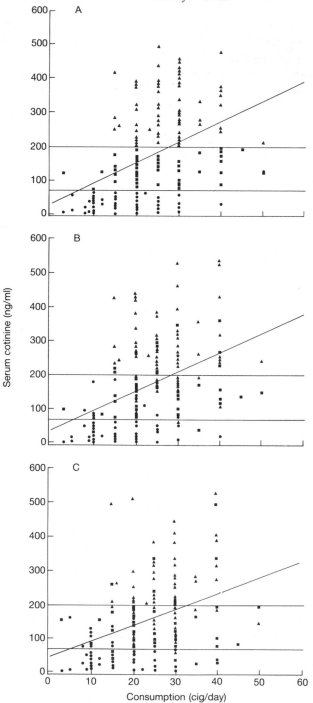

Fig. 9.2. Relationship between serum cotinine level and number of cigarettes smoked per day (Study 1). (A) 1st investigation: $n = 158$; $r = 0.44$. (B) 2nd investigation: $n = 158$; $r = 0.44$. (C) 3rd investigation: $n = 157$; $r = 0.41$. ●, <70 ng cotinine/ml at 1st investigation; ■, 70–200 ng cotinine/ml at 1st investigation; ▲, >200 ng cotinine/ml at 1st investigation.

while only nine out of 158 subjects (6 per cent) cross both borderlines. Between-smoker standard deviations for serum nicotine and cotinine were about twice the within-smoker standard deviation (Table 9.1). The mean nicotine yields of the cigarettes smoked in the low, medium, and high cotinine group were 0.7 (range 0.4–1.5), 0.9 (0.4–1.5), and 1.0 (0.5–1.7) mg per cigarette, respectively, while their reported average daily cigarette consumptions were 16.4 (range 3–40), 24.1 (3–50), and 27.9 (15–50), respectively. Smokers in the high and medium cotinine groups tended to have had a longer smoking career and reported inhaling more deeply than those in the low cotinine group.

Table 9.1. Inter- and intra-individual variation in smokers' serum nicotine and cotinine (Study 1)

	SD between smokers			SD within smokers
	Time 1	Time 2	Time 3	
Nicotine (ng/ml)	11.7	11.8	12.2	6.4
Cotinine (ng/ml)	136.1	133.6	125.2	62.6

Study 2

Figure 9.3A (males) and 9.3B (females) show the range of values of serum cotinine, cigarette consumption, and nicotine delivery (= cigarette consumption × nicotine yield/cigarette) observed in each individual, ranked according to their mean cotinine level. Mean cotinine levels varied from about 10 ng/ml to over 800 ng/ml and were clearly higher in men (mean 275.1 ng/ml) than in women (mean 214.2 ng/ml, $P < 0.05$). Figure 9.3 shows that in both sexes between-smoker variability is greater than within-smoker variability for all three parameters. This is quantified in Table 9.2 which also includes results for serum nicotine. Generally, between-smoker standard deviations were about twice as high as within-smoker standard deviations. Between-smoker standard deviations were higher in men than in women for serum nicotine and cotinine, although not for nicotine delivery or cigarette consumption. For the six points of time correlation coefficients for cotinine against cigarette consumption ranged from 0.33 to 0.51, while those for cotinine against nicotine delivery ranged from 0.12 to 0.39.

Model 1 (see section 'Statistical methods') explained a highly significant ($P < 0.001$) 85.6 per cent of the variation in serum cotinine, thus confirming the large between-person variability. Model 2 explained a further highly significant ($P < 0.001$) 2.3 per cent of the variation, with the compensation index in relation to number of cigarettes smoked estimated as $\alpha = 0.587$ (SE = 0.070). The introduction of nicotine yield (Model 3) proved of

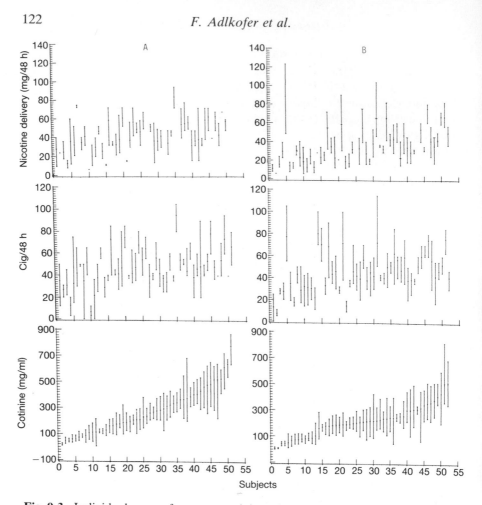

Fig. 9.3. Individual ranges for serum cotinine, cigarette consumption and nicotine delivery, ranked according to the mean cotinine levels (Study 2). (A) Males. (B) Females.

marginal significance ($P < 0.05$), with the compensation index estimated as $\beta = 0.245$ (SE $= 0.120$), α being hardly changed at 0.593 (SE $= 0.069$). The higher variability for β is due to many smokers showing little or no change in nicotine yield over the six points of time.

The assumption that α and β were constant from person to person was tested by additional models in which α and β were replaced by α_i and β_i. These did in fact show some improvement in fit for both the cigarette consumption index ($P < 0.001$) and for nicotine yield index ($P < 0.001$).

Further analysis showed that the compensation index in relation to number of cigarettes smoked (α) was not significantly related to sex (males:

Table 9.2. Inter- and intra-individual variation in smokers' serum nicotine and cotinine as well as in cigarette consumption and nicotine delivery (Study 2)

| | Sex | SD between smokers | | | | | | SD within smokers |
		Time 1	Time 2	Time 3	Time 4	Time 5	Time 6	
Nicotine (ng/ml)	M	10.4	18.8	15.9	12.9	12.7	13.5	8.5
	F	10.2	11.3	9.5	11.3	10.9	11.1	7.1
Cotinine (ng/ml)	M	184.0	185.2	190.7	197.1	177.3	193.3	63.6
	F	135.5	137.9	131.6	129.0	162.7	121.3	63.3
Consumption (cig/48h)	M	18.1	19.0	16.1	20.6	25.7	19.4	9.4
	F	18.8	19.8	19.9	21.0	18.6	22.4	9.3
Nicotine delivery (mg/48h)	M	18.6	18.3	16.6	20.4	18.2	19.8	9.1
	F	20.2	19.9	21.1	20.7	28.1	24.3	9.4

$\alpha = 0.48$, SE $= 0.12$; females: $\alpha = 0.65$, SE $= 0.09$), to normal cigarette consumption ($\alpha = 0.72$, 0.55, 0.47, 0.40, and 0.69 for smokers of <20, 20, 21–29, 30, or >30 cigarettes/day, respectively) or to nicotine yield of brand normally smoked ($\alpha = 0.53$, 0.61, and 0.62 for smokers of <0.8, 0.8, and >0.8 mg/cigarette, respectively). There was also no evidence of non-linearity, i.e. it seems that the overall estimate of $\alpha = 0.593$ gives a reasonable indication of the average extent to which a smoker's serum cotinine will be affected by changes in cigarette consumption.

The compensation index in relation to nicotine yield (β) was not significantly related to sex (males: $\beta = 0.30$, SE $= 0.22$; females: $\beta = 0.23$, SE $= 0.14$). Nor was it related to normal cigarette consumption, with no evidence of non-linearity seen. However, there was a tendency for estimates of β to be lower in smokers whose normal brand had a nicotine yield of <0.8 mg/cigarette ($\beta = 0.15$, SE $= 0.22$) or 0.8 mg/cigarette ($\beta = 0.09$, SE $= 0.15$) as compared to those whose normal brand had a nicotine yield of >0.8 mg/cigarette ($\beta = 1.36$, SE $= 0.36$, $P < 0.001$).

As well as showing the within-smoker estimates of α and β, Table 9.3 indicates between-smoker estimates for each of the six points of time.

Table 9.3. Estimates of compensation indexes (Study 2)

	Cigarette consumption α Mean (SE)	Nicotine yield β Mean (SE)
Within-smoker	0.593 (0.069)	0.245 (0.120)
Between-smoker		
Time 1	1.27 (0.20)	0.32 (0.27)
Time 2	0.84 (0.18)	0.50 (0.24)
Time 3	0.91 (0.20)	0.46 (0.24)
Time 4	0.70 (0.20)	0.34 (0.24)
Time 5	0.97 (0.21)	0.30 (0.24)
Time 6	0.91 (0.19)	0.48 (0.24)

Between-smoker estimates tend to be higher, averaging 0.93 and 0.43, respectively. For reasons discussed later, we assume that these estimates are probably biased upwards.

9.4. Discussion

The question of compensation is of vital importance for the future development of safer cigarettes. In the past, two main approaches have been used to try to determine whether smokers compensate for any changes in cigarette yields as measured by smoking machine.

1. Cross-sectional field studies in which smokers are allowed to smoke their own brand freely and in which smoke intake (usually measured as nicotine or cotinine level in body fluids) at one point in time is related to the number of cigarettes smoked and the true nicotine yield of the brand smoked.[9-13]

2. Controlled laboratory experiments in which the subjects smoke a series of cigarettes varying in their machine-smoked yields, with smoke intake again being related to the number and yield of the cigarettes smoked.[14-17]

The first approach is on a between-subject basis, the second on a within-subject basis.

Both approaches have a number of advantages and disadvantages. The field studies are usually based on large numbers of subjects and the smokers smoke freely in their everyday environment. However, owing to their cross-sectional character, these studies tell us nothing about within-smoker variation and do not allow direct inference about changes in smoke intake in relation to changes in smoking habits. The controlled laboratory studies do give us information on within-smoker variation, but suffer from the 'artificial' smoking situation and the limited numbers of subjects observed (usually less than 20). Moreover, due to the relatively short observation periods, the smokers do not have time to adjust completely to the new brands.

We tried to overcome some of these disadvantages by carrying out longitudinal field studies. To our knowledge, apart from a few observations,[3-5] the problem of compensation has not been approached by such a study design.

The principal objective of our studies was to find out whether individual smokers take in constant amounts of nicotine over time. In our view, consistency of nicotine intake would provide further evidence for regulation of nicotine intake. The results of our first study suggest that despite quite a large variation in serum cotinine levels between individuals, cotinine remains relatively constant over time for a considerable number of smokers. Relatively weak associations of serum cotinine with either cigarette consumption or the nicotine yield of the cigarettes smoked were found. This is in agreement with the results of cross-sectional studies performed by several other groups.[9-13] However, there were indications that compensation was unlikely to be complete, since the groups with high and medium cotinine levels tended to smoke more, to smoke higher yield cigarettes, to have smoked for longer and to report deeper inhalation as compared to the low cotinine group. This is in line with the conclusions of an IARC working group.[1]

One difficulty encountered in our first study was that data were not collected on cigarette consumption and brand nicotine yield at each point of time. The relationship of these parameters to cotinine levels could therefore be studied only on a between-smoker basis. Where such large inter-individual differences in cotinine levels for a given nicotine intake are present, these relationships might well differ systematically from those which would be observed on a within-smoker basis.

In order to base our conclusions on a broader set of data, we performed a second study which was extended to both sexes and to six observations per individual over time, with cigarette consumption and brand nicotine yield measured on each occasion. A mathematical compensation model was applied to the data of Study 2 which allows quantification of the relationship of serum cotinine levels to cigarette consumption and to nicotine yield. For two reasons the main modelling work was carried out on a within-smoker basis. Firstly, it was already clear from Fig. 9.3 that a simple version of the model in which a single α, β, and log μ was fitted could not explain the between-smoker variation, since subjects with similar cigarette consumption and nicotine yield can be seen to have consistently very different cotinine values. Secondly, between-smoker estimates of α and β may be biased upwards if there is a positive correlation between inhalation and number of cigarettes or nicotine yield. The model quantitatively reveals what is qualitatively evident from a visual inspection of Figs 9.2 and 9.3; the major part of the variation in the serum cotinine (85.6 per cent) is explained by the large between-person variability. This is consistent with the existence of substantial differences between the way in which people inhale[18] or metabolize nicotine.[19] Less important, but still significant in explaining variation in serum cotinine are the number of cigarettes smoked, and the brand nicotine yield. The magnitudes of the compensation indices ($\alpha = 0.593$, $\beta = 0.245$) clearly demonstrate partial compensation. The values mean that halving cigarette consumption should reduce cotinine concentration by 34 per cent, whereas halving nicotine yield should produce a 16 per cent reduction. The former estimate confirms recent results obtained by Benowitz *et al.*[20] The model also implies that the compensation indices do not significantly depend on sex or cigarette consumption, although the data suggest that halving nicotine yield will produce a greater reduction for higher-yield cigarettes than for lower-yield cigarettes.

We are well aware that our study design and compensation model may have certain weaknesses. One is clearly the fact that most of our subjects did not vary consumption or yield dramatically over the six points of time. This could increase the error in estimating the compensation indices and may bias them downwards in our within-person approach. To rectify this one would need a study in which the time points are much further apart. Another weakness might be that our compensation model assumes that all smokers compensate to the same extent for a given reduction in number of cigarettes or cigarette yield. It seems more likely that smokers may differ in their behaviour according to their motivation for smoking and their reasons for changing brands. It is interesting to note that compensation was more complete for smokers of lower yield cigarettes. This is consistent with the idea that smokers might tend to compensate if their nicotine level were to drop below some required minimum. A more general deficiency of our compensation model is related to the group of smokers with very low serum cotinine levels. These smokers might not smoke for nicotine since very low nicotine doses are unlikely to exhibit pharmacological effects. The low compensation indices as measured in our model would falsely indicate considerable or complete compensation. Of course, it must be borne in mind that our study does not provide direct evidence that it is nicotine rather than another substance in tobacco smoke for which smokers compensate. However, due to its neuropharmacological effects, nicotine seems most likely to determine the extent of smoke intake. A number of studies, including some very recent ones,[21-23] provide direct evidence that smokers regulate their nicotine intake.[2,24]

The data so far available, including our own, justify the assumption that a decrease in the tar/nicotine ratio in cigarettes might reduce the intake of toxic substances, thus diminishing the health risk due to smoking. Since the nicotine yield per cigarette need not be increased beyond the present average level, this assumption holds true even if nicotine is involved in the pathogenesis of any smoking-related disease, for which in our view there is no convincing evidence. The absence of an increased risk of coronary heart disease in pipe and cigar smokers, even though their nicotine intake is relatively high,[25,26] may be regarded as evidence against a significant role of nicotine in the pathogenesis of this disease. A more serious problem may arise from the fact that the alkaloid is a precursor of tobacco-specific nitrosamines (TSNA).[27] Whether the TSNA are more important in smoking-related cancerogenesis than other constituents of tobacco smoke is not yet known. Even if TSNA are involved, this should not affect the suggested concept of the less harmful cigarette. On the other hand, there are ample reasons for developing technological procedures to reduce the TSNA content in the mainstream smoke of cigarettes. Initial steps are already being taken in this direction.

If it is assumed that smoking has redeeming values and that many smokers are therefore not willing to give it up, we have to decide whether, apart from cessation and education programmes, a product modification programme towards safer cigarettes would have a chance. In our opinion, the following considerations are of importance for the development of a safer cigarette in the future.

1. The tar/nicotine ratio should be kept as low as possible, thus leading to

a reduced intake of total particulate matter. Trends in this direction have already been set, particularly in UK, but also in the USA and West Germany.

2. The lowest possible tar yield of such cigarettes should be established. The limitation of this development should be seen in the acceptance by the consumer. The safest cigarette would be useless if rejected by the smoker.

3. The alteration of the tar/nicotine ratio should be accompanied by a selective lowering or removal of substances in tobacco smoke which are identified as toxic or carcinogenic.

4. Further investigation is needed to find out whether or not nicotine is involved in the pathogenesis of smoking-related diseases. Since nicotine is essential for the smoker it cannot be eliminated from cigarette smoke.

Acknowledgements

We thank Mrs C. Hrubý, Mrs U. Stoltze, and Dr G. Henze for their invaluable assistance in preparing the manuscript. The views taken in this paper express the personal opinion of the authors.

References

1. International Agency for Research on Cancer (IARC) (1986). *Tobacco Smoking*. IARC Monographs on the evaluation of the carcinogenic risk of chemicals to humans, Vol. 38, pp. 170–9. IARC, Lyon, France.
2. McMorrow, M. J., and Foxx, R. M. (1983). Nicotine's role in smoking: An analysis of nicotine regulation. *Psychological Bulletin*, **93**, 302–27.
3. Haley, N. J., Axelrad, C. M., and Tilton, K. A. (1983). Validation of self-reported smoking behaviour: Biochemical analyses of cotinine and thiocyanate. *American Journal of Public Health*, **73**, 1204–7.
4. Gori, G. B. and Lynch, C. J. (1983). Smoker intake from cigarettes in the 1-mg Federal Trade Commission tar class. *Regulatory Toxicology and Pharmacology*, **3**, 110–20.
5. Robinson, J. C., Young, J. C., and Rickert, W. S. (1984). Maintain levels of nicotine but reduce other smoke constituents: A formula for 'less hazardous' cigarettes? *Preventive Medicine*, **13**, 437–45.
6. Hengen, N. and Hengen, M. (1978). Gas-liquid chromatographic determination of nicotine and cotinine in plasma. *Clinical Chemistry*, **24**, 50–3.
7. Langone, J., Cjika, H. B., and Van Vunakis, H. (1973). Nicotine and its metabolites: Radioimmunoassays for nicotine and cotinine. *Biochemistry*, **12**, 1092–8.
8. Biber, A., *et al.* (1987). Determination of nicotine and cotinine in human serum and urine – An interlaboratory study. *Toxicology Letters*, **35**, 45–52.

9. Russell, M. A. H., Jarvis, M., Iyer, R., and Feyerabend, C. (1980). Relation of nicotine yield of cigarettes to blood nicotine concentrations in smokers. *British Medical Journal*, **280**, 972–6.
10. Rickert, W. S. and Robinson, J. C. (1981). Estimating the hazards of less hazardous cigarettes. II. Study of cigarette yields of nicotine, carbon monoxide, and hydrogen cyanide in relation to levels of cotinine, carboxyhemoglobin and thiocyanate in smokers. *Journal of Toxicology and Environmental Health*, **7**, 391–403.
11. Benowitz, N. L., Hall, S. M., Herning, R. I., Jacob, P., III, Jones, R. T., and Osman, A. L. (1983). Smokers of low-yield cigarettes do not consume less nicotine. *New England Journal of Medicine*, **309**, 139–42.
12. Hill, P., Haley, N. J., and Wynder, E. L. (1983). Cigarette smoking: Carboxyhemoglobin, plasma nicotine, cotinine and thiocyanate vs self-reported smoking data and cardiovascular disease. *Journal of Chronic Diseases*, **36**, 439–49.
13. Gori, G. B., and Lynch, C. J. (1985). Analytical cigarette yields as predictors of smoke bioavailability. *Regulatory Toxicology and Pharmacology*, **5**, 314–26.
14. Ashton, H., Stepney, R., and Thompson, J. W. (1979). Self titration by cigarette smokers. *British Medical Journal*, **2**, 357–60.
15. Hill, P. and Marquardt, H. (1980). Plasma and urine changes after smoking different brands of cigarettes. *Clinical Pharmacology and Therapeutics*, **27**, 652–8.
16. Stepney, R. (1981). Would a medium-nicotine, low-tar cigarette be less hazardous to health? *British Medical Journal*, **283**, 1292–6.
17. Russell, M. A. H., Sutton, S. R., Iyer, R., Feyerabend, C., and Vesey, C. J. (1982). Long term switching to low-tar low-nicotine cigarettes. *British Journal of Addiction*, **77**, 145–58.
18. Herning, R. I., Jones, R. T., Benowitz, N. L., and Mines, A. H. (1983). How a cigarette is smoked determines blood nicotine levels. *Clinical Pharmacology and Therapeutics*, **33**, 84–9.
19. Benowitz, N. L., Jacob, P., III, Jones, R. T., and Rosenberg, J. (1982). Interindividual variability in the metabolism and cardiovascular effects of nicotine in man. *Journal of Pharmacology and Experimental Therapeutics*, **221**, 368–72.
20. Benowitz, N. L., and Jacob, P., III, Kozlowski, L. T., and Yu, L. (1986). Influence of smoking fewer cigarettes on exposure to tar, nicotine, and carbon monoxide. *New England Journal of Medicine*, **315**, 1310–3.
21. Russell, M. A. H. (1986). *The role of nicotine in compensatory smoking*. Paper presented at the Independent Scientific Committee on Smoking and Health Symposium 'Nicotine, smoking and the low tar programme', 18–20 November 1986, London. (This volume, chapter 11.)
22. Herning, R. I., Jones, R. T., and Fishman, P. (1985). The titration hypothesis revisited: Nicotine gum reduces smoking intensity. In: *Pharmacological aspects in smoking cessation* (ed. J. Grobowski and M. H. Sharon). NIDA Research Monograph, **53**, 27–41.
23. Pomerleau, C. S., Pomerleau, O. F., and Majchrzak, M. J. (1987). Mecamylamine pre-treatment increases subsequent nicotine self-administration as indicated by changes in plasma nicotine level. *Psychopharmacology*, **91**, 391–3.
24. Moss, R. A. and Prue, D. M. (1982). Research on nicotine regulation. *Behaviour Therapy*, **13**, 31–46.

25. Wald, N. J., Idle, M., Boreham, J., Bailey, A., and Van Vunakis, H. (1981). Serum cotinine levels in pipe smokers: Evidence against nicotine as cause of coronary heart disease. *Lancet*, **ii,** 775–7.
26. Schievelbein, H., Adlkofer, F., Scherer, G., Biber, A., and Heller, W. D. Nicotine uptake by pipe, cigar and cigarette smokers (in preparation).
27. Hoffmann, D., Lavoie, E. J., and Hecht, S. S. (1985). Nicotine: A precursor for carcinogens. *Cancer Letters*, **26,** 67–75.

IV

Determinants of Low Tar Smoking and of Compensation

10

Dosimetric studies of compensatory cigarette smoking

NEAL L. BENOWITZ

Abstract

Methods for estimating tar, nicotine, and carbon monoxide exposure are reviewed and their limitations noted. Results from experimental and spontaneous brand switching studies have shown only small reductions in exposure to tobacco smoke toxins when switching from high or middle to low-yield cigarettes; the reduction was slightly greater when switching to ultra-low yield products. Switching to a higher-yielding cigarette should be discouraged since data indicate the possibility of greater exposure to tobacco toxins.

10.1. Introduction

At issue at this time is whether some modern cigarettes are less hazardous than others and whether public health policy should be to encourage people who cannot stop smoking to switch to low-yield cigarettes. Because of the latency in development of most smoking-related diseases, direct assessment of the health risks of low-yield cigarettes is impossible. An adequate epidemiological study would take 20 or more years.

10.2. Biological dosimetry

Indirect assessments of health risk may include measurements of intermediary biological effects. Considering smoking-related cancer, this might include chromosomal abnormalities or abnormal sister-chromatid exchange in lymphocytes. The latter has been described in cigarette smokers,[1] but the sensitivity of this test is inadequate for quantitative risk estimation. If biological effects cannot be monitored, the most rapid and direct way to compare risks of different cigarettes is to measure human exposure to chemicals in tobacco smoke which are believed to cause disease.

Estimating tar exposure

In studying tobacco-related carcinogenesis, polycyclic aromatic hydro-carbons, various nitrosoamines, naphthylamines, and polonium-210 have been of particular interest. Unfortunately, no direct assays for these sub-stances in human biological fluids are currently available. A crude approach to estimating exposure to potentially carcinogenic chemicals is the measurement of mutagenic activity of the urine. This is commonly done using the Salmonella histidine auxotroph reversion assay.[2] *In vitro* studies indicate that the mutagenic components of cigarette smoke are found primarily in the tar rather than the gaseous fraction.[3] It is known that the urine of cigarette smokers is mutagenic. For an individual, muta-genic activity of the urine tends to be constant from day to day (Fig. 10.1) and there is a relationship between mutagenic activity and the number of cigarettes smoked per day.[4–6] The test is limited in that it is not specific for exposure to particular carcinogens, there is considerable variability in results from assay to assay and from person to person, and dietary and environmental chemical exposures can influence mutagenic activity. How-ever, for within-subject comparisons, when assays are performed in one run and compared for the same individual, the test seems to provide a quantitative estimate of exposure to tar and, hence, potential carcinogen exposure.

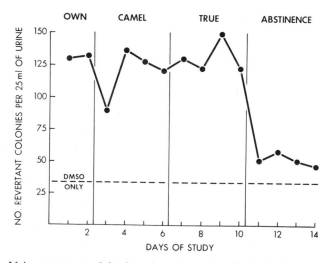

Fig. 10.1. Urinary mutagenicity based on 24-hour urine collections in an habitual smoker while smoking own brand, Camel (high-yield) and True (low-yield) cigar-ettes. Note that mutagenic activity tends to be constant from day to day and falls to the dimethyl sulphoxide (DMSO) control value (similar to that of nonsmokers) rapidly after stopping smoking.

Estimating nicotine and carbon monoxide exposure

Nicotine can be measured directly in human blood and other biological fluids. Carbon monoxide can be measured in blood and expired air. The role of nicotine and carbon monoxide in causing human disease is not proven, but is suspected for cardiovascular disease and reproductive disorders. Nicotine is also a useful marker for tar exposure. In smoking machine tests, deliveries of nicotine and tar are generally correlated.[7] There are differences in machine-determined tar-to-nicotine ratios from brand to brand, the ratios decreasing as overall yield decreases. An attempt has been made to adjust for these differences in using the machine-derived tar-to-nicotine ratio in combination with nicotine exposure measurements to generate a tar exposure index.[8] A limitation in such an estimate is that the tar-to-nicotine ratio may change when cigarettes are smoked in a manner different from that of the smoking machine. This issue will be discussed in further detail later in this paper.

Measuring daily intake of nicotine in cigarette smokers

Exposure to nicotine has been estimated directly by measuring blood concentrations of nicotine in smokers. Because of the relatively short half-life of nicotine (2 hours)[9] and considerable variation in nicotine levels according to time of day and time from last cigarette, single measurements of nicotine levels are generally thought not to reflect total daily exposure. A better way to measure exposure is to sample nicotine levels frequently during the day. Daily exposure can then be computed as a time-weighted average, operationally the area under the blood nicotine concentration-time curve (AUC_{nic}). Similar measurements can be made to estimate daily carbon monoxide exposure (AUC_{COhB}).

However, because of individual differences in the rate of metabolism of nicotine,[9] exposure does not directly indicate *intake* of nicotine. That is, with the same level of intake fast metabolizers will have lower nicotine concentrations than slow metabolizers. The relationship is given by the equation: daily intake = clearance $\times$ AUC_{nic}. Clearance is a mathematical term which relates the rate of elimination and the blood concentration of a drug. It is the term which best describes the capacity of a person to metabolize and/or excrete a drug. Using clearance estimates derived from intravenous nicotine infusion studies combined with AUC measurements during *ad libitum* smoking, daily intake of nicotine can be measured.[10] Such studies indicate a range of intake from 8 to 80 mg per day. Average intake of nicotine per cigarette was 1.0 mg, but ranged from 0.4 to 1.6 mg. As expected from known differences between how people and machines smoke cigarettes, there was not a consistent relationship between machine yield and actual intake of nicotine per cigarette.

For studies of larger populations, it is desirable to have a biological marker which can be measured only once in a day. Cotinine, the major metabolite of nicotine, has a long half-life (averaging 19 hours) and cotinine levels vary much less through the day than do levels of nicotine (Fig. 10.2).[11] For this reason, cotinine has become the preferred marker for intake of nicotine. However, the level of cotinine does not accurately describe intake of nicotine. Individual differences in the proportion of nicotine converted to cotinine and in the rate of metabolism of cotinine *per se* influence the proportionality between nicotine and cotinine levels as described in the equation:

$$D_{nic} = \frac{BC_{cot} \times CL_{cot}}{F}$$

where D_{nic} is daily nicotine intake, BC_{cot} is the steady state (or average) blood concentration of cotinine, CL_{cot} is clearance of cotinine, and F is

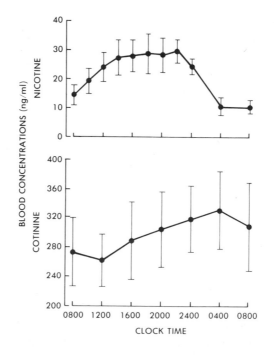

Fig. 10.2. Circadian blood concentrations of nicotine and cotinine during unrestricted smoking. Although fluctuating far less than nicotine, cotinine concentrations vary by 30 per cent throughout the day. Data are shown as mean ± SE for eight subjects (from Benowitz *et al.*[11]).

fractional conversion of nicotine to cotinine. Additionally, although cotinine levels are more constant than are levels of nicotine throughout the day, cotinine levels do vary from morning to evening with regular smoking (Fig. 10.2). Therefore, to optimize its usefulness for comparative studies, time of day of cotinine sampling needs to be considered and standardized.

Using daily intake of nicotine estimated from 24-hour blood level sampling and intravenous nicotine clearance, the usefulness of various biological markers has been examined (Table 10.1). In the conditions of this study, that is, sedentary smokers, on a research ward, free to smoke at any time, casual (not timed in relationship to the smoking of any cigarette) blood

Table 10.1. Correlation of various markers of tobacco smoke intake with daily intake of nicotine and nicotine exposure

Measure	Time	Intake	AUC
BNC	4 p.m.	0.81*	0.91*
BNC	Noon	0.76*	0.90*
COHB	4 p.m.	0.69*	0.81*
COHB	Noon	0.65*	0.82*
U_{cot}	24 hr	0.62*	0.46†
COHB	8 a.m.	0.61*	0.76*
BNC	8 a.m.	0.56*	0.78*
B_{cot}	4 p.m.	0.53†	0.45†
B_{cot}	8 a.m.	0.51†	0.39
U_{nic}	24 hr	0.39	0.31

BNC = blood nicotine concentration; COHB = carboxyhaemoglobin level; U_{cot} = 24-hr excretion of cotinine; B_{cot} = blood cotinine concentration.
*$P < 0.001$; †$P < 0.05$.

levels of nicotine and carboxyhaemoglobin measured in the afternoon (when levels of nicotine and carbon monoxide plateau during regular smoking) best predicted daily nicotine intake. Blood or urinary cotinine correlated less well, with correlation coefficients of only 0.5–0.6. Because of individual differences in metabolism, this may be the best achievable degree of correlation. This degree of correlation must be considered in interpreting studies in which cotinine is taken as the quantitative marker of nicotine intake.

10.3. Studies of compensatory smoking

As has been discussed in many publications and in other chapters within this volume, people smoke cigarettes differently to how smoking machines do.

They tend to regulate intake of nicotine to maintain, within certain limits, a particular level of nicotine in the body. Issues which need to be resolved are how complete is that compensation and is the extent of compensation the same for all low-yield cigarettes? Such issues have been addressed by several types of research studies. First, there are experimental studies in which smokers are switched to different brands of cigarettes for the purposes of the study. Biochemical markers of tobacco smoke exposure are measured while smoking usual brands and again after switching. A second type of experiment is the population study in which the relationship between brands of cigarettes selected by the smoker and biochemical measures of exposure to smoke components are examined. A third type of study is one in which the natural history of brand switching is examined. That is, subjects are followed over time and biochemical measurements are made before and after spontaneous changes in brands. Examples of these three types of studies from our laboratory will be discussed. Other studies of compensatory smoking are described by other authors in this volume.

Experimental brand switching studies

Our laboratory sought to determine if exposure to cigarette toxins is reduced by switching to modern low- or ultra-low-yield American cigarettes. [12,13] Definitions of low-yield cigarettes have changes over the years. For purposes of our studies, cigarettes of different nicotine yield have been defined as follows: high − greater than 1.00 mg nicotine; medium − 0.61–1.00 mg; low − 0.20–0.60 mg; and ultra-low − less than 0.20 mg.

Subjects were hospitalized on a research ward for the duration of the study. Two groups of subjects were studied: one was switched from their usual brand of cigarette to a high-yield and a low-yield cigarette; the other group was switched to a high-yield and an ultra-low-yield cigarette. Yields for the usual brand of cigarettes and test cigarettes are shown in Table 10.2. Each cigarette was smoked for four days. The sequence of cigarette brands was balanced. Exposure to nicotine and carbon monoxide was measured by frequent blood sampling and expressed as area under the blood concentration time curve (*AUC*). Tar exposure was estimated as mutagenic activity of 24-hour urine collections.

When switched from their own brand to a high-yield brand, which was of similar yield to their own selected brand, intake of nicotine and other measures decreased by about one-third (Fig. 10.3). Thus, the high-yield test cigarettes were smoked differently from the subject's own brand. This is probably because of differences in taste or other characteristics such that the cigarettes were not as pleasurable as their own brand and were smoked with less enthusiasm. Changes in intake when switching cigarettes, even of the same nominal yield, must be considered in evaluating the results of other experimental studies.

Table 10.2. US Federal Trade Commission smoking machine yields of usual and test brands of cigarettes*

	Tar (mg)	Nicotine (mg)	Carbon monoxide (mg)	Tar/nicotine ratio	CO/nicotine ratio
Group 1					
Usual†	16.3 ± 2.7	1.10 ± 0.15	15.1 ± 1.3	14.7 ± 1.0	14.2 ± 0.6
High	15.4	1.0	15.2	15.4	15.2
Low	4.6	0.4	4.8	11.5	12.0
Group 2					
Usual	14.7 ± 4.6	1.06 ± 0.32	14.1 ± 4.0	13.6 ± 1.9	13.4 ± 2.1
High	15.4	1.0	15.2	15.4	15.2
Ultra-low	0.8	0.1	1.4	8.0	14.0

*Based on 1983 FTC report.[30]
†Usual brand, mean ± SD.

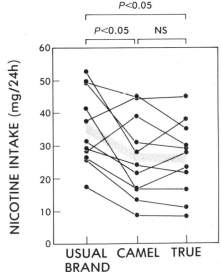

Fig. 10.3. Daily intake of nicotine while smoking usual brand, Camel or True cigarettes. Machine-determined yields are given in Table 10.2 (from Benowitz and Jacob[13]).

Exposures from low- and ultra-low-yield cigarettes were, therefore, compared to exposures when switched to high-yield cigarettes (Fig. 10.4). When switched to low-yield cigarettes, subjects reported these cigarettes to be less satisfying than their usual brand. They smoked more cigarettes, but the amount of tobacco burned was unchanged. This is because low-yield cigarettes contain less tobacco per cigarette.[14] However, intake of nicotine,

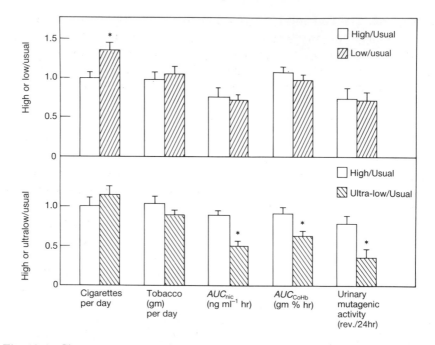

Fig. 10.4. Cigarette consumption, nicotine (AUC_{nic}) and carbon - monoxide (AUC_{CoHb}) exposure, and urinary mutagenicity while smoking different brands of cigarettes. Data are expressed as average ($N = 11$) ratio of values while smoking test high-, low-, or ultra-low-yield cigarettes to values while smoking usual brand. Machine-determined yields are given in Table 10.2. Bars indicate SEM; asterisk indicates significant differences ($P<0.05$) between low- or ultral-low-, and high-yield cigarettes; rev indicates revertant colonies (from Benowitz *et al.*[12]).

carbon monoxide, and urinary mutagenicity were identical comparing low- and high-yield cigarettes, indicating full compensation. Of note is the high concordance ($r = 0.86$) between nicotine intake for individuals comparing the high- and low-yield conditions (Fig. 10.3). This finding supports the concept of regulation of nicotine intake for individuals

Subjects found the ultra-low-yield cigarettes to be unsatisfying, often describing them to be like 'smoking air'. Exposures to nicotine, carbon monoxide, and tar were reduced by 56, 36, and 49 per cent, respectively, compared with the high-yield test cigarette. Although exposure was reduced compared with other cigarettes, smokers of ultra-low-yield cigarettes still consumed far more tobacco smoke than predicted by the smoking machine determined yield. Ultra-low-yield cigarettes have such low smoking machine deliveries primarily because of extensive ventilation, resulting in dilution of the smoke by air. The smoker can increase yield substantially by obstructing ventilation holes in the filter, often done unknowingly because it improves the draw characteristics of the cigarette.[15-18] However, our smokers were unable to fully compensate. This is presumably because the ventilation holes can only be partially obstructed by the smoker. Two other recent studies support the conclusion that toxic exposures from ultra-low yield cigarettes differ from exposures from other cigarettes.[19,20]

Tar- and carbon monoxide-to-nicotine ratios

Smoking machine tests suggest that low-yield cigarettes will have a lower tar-to-nicotine ratio[21] (also see Table 10.2). Based on smoking machine data, it has been predicted that even if nicotine intake is regulated, tar exposure, and therefore risk of cancer, will be reduced.[21] Our data allowed us to compare tar-to-nicotine ratios predicted by the machine with that observed in human smokers. We found that the ratio of urine mutagenicity to AUC_{nic} was similar for all brands (Fig. 10.5). There is a slight tendency for a lower ratio for the ultra-low-yield brands, but if so, the ratio is still far greater than the 50 per cent reduction in ratio predicted by smoking machine tests. Smoking machine studies indicate that hole-blocking and/or

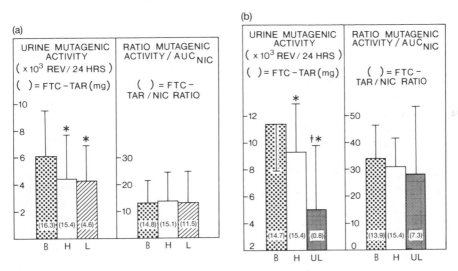

Fig. 10.5. Average urine mutagenicity and ratio of mutagenic activity to nicotine exposure (AUC_{nic}) for subject in high–low-yield and high–ultra-low-yield comparison studies. Note that despite lower ratios of tar/nicotine based on United States Federal Trade Commission testing for low- and ultra-low-yield cigarettes, ratios of mutagenic activity to nicotine exposure were not different while smoking high (H)-, low (L)-, or ultra-low (UL)-yield cigarettes. Bars indicate SEM; asterisk indicates significant differences compared with usual-brand baseline (B); double dagger indicates significant difference compared with high-yield cigarettes (from Benowitz *et al.*[12]).

more intensive smoking results in increased tar-to-nicotine ratios, especially for lower yield cigarettes[17,18,22] (Table 10.3). Based on these data, we question whether predictions about safety of different brands of cigarettes based on machine determined tar-to-nicotine ratios or estimates of tar intake based on measures of nicotine intake plus machine derived tar-to-nicotine ratios are valid.

Table 10.3. Influence of intensity of smoking on tar/nicotine ratio, based on smoking machine studies†

Group	N	Standard yield		Tar/nicotine ratio under different smoking conditions		
		Tar (mg)	Nicotine (mg)	Standard	Moderate	Intensive
I	4	<2	<2	9.2	9.9	11.1*
II	10	2–5	0.2–0.5	10.3	11.7*	12.2*
III	8	5–10	0.5–0.9	11.3	11.9	12.6*
IV	9	10–14	0.8–1.0	12.7	13.3*	12.4
V	5	14–17	0.9–1.0	15.7	16.5*	14.7

*$P < 0.05$ compared to STD.
†Rickert et al.[22]
N, number of brands tested.

Ratios of carbon monoxide to nicotine are predicted by machine to be slightly lower for low-yield cigarettes and similar for high- and ultra-low-yield cigarettes (Table 10.2). We found no difference in AUC_{CO}/AUC_{nic} ratios for high- and low-yield cigarettes (0.65 ± 0.23 v. 0.58 ± 0.16 per cent/ng/ml) and a significantly higher ratio for ultra-low-yield cigarettes (0.58 ± 0.17 v. 0.75 ± 0.22 per cent/ng/ml, $P < 0.05$). Other studies have also noted a lesser reduction of carbon monoxide than nicotine when switching to ultra-low-yield cigarettes.[23,24] In general, other gaseous compounds are generated in proportion to carbon monoxide.[25] Our data suggest that exposure to gaseous toxins will be minimally reduced by switching even to the ultra-low-yield cigarettes.

Population studies

Experimental brand switching studies are artificial in that smokers change cigarette brands only for the purpose of the research; motivation and cigarette acceptability are dissimilar to the natural situation of brand switching; and studies are conducted over relatively brief periods of time. Population studies provide data as to exposure to tobacco smoke toxins of people who have selected brands they find satisfying.

A group of 248 subjects was studied.[12] These were subjects about to start a smoking treatment programme. The group consisted of 116 men and 132 women with an average age of 37 years (range 22–55 years). The subjects smoked an average of 30 (range 4–100) cigarettes per day with an average nicotine yield of 0.75 ± 0.35 (SD) mg nicotine. Blood samples for analysis of nicotine, cotinine, and thiocyanate were obtained, and expired carbon monoxide levels were measured late in the afternoon. As seen in Fig. 10.6, blood levels of nicotine, cotinine, and thiocyanate were 27, 39, and 27 per cent lower, respectively, for subjects smoking ultra-low, but no different for subjects smoking higher yield cigarettes. Carbon monoxide exposure was similar for all brands.

Our data indicate little difference in tobacco smoke exposure among all but the ultra-low-yield cigarette smokers. Our sample was limited by relatively small size and the fact that these were all habitual smokers who were entering a smoking cessation programme. Other larger populations of smokers indicate that there may be a continuous linear relationship between nicotine yield and intake, but that the slope of the line is quite shallow (Fig. 7).[8,26,27] Thus, across the range of cigarette brands there may

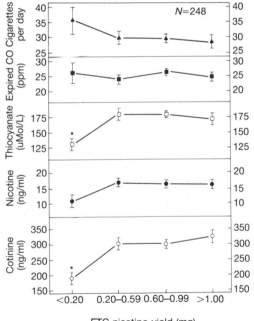

FTC nicotine yield (mg)

Fig. 10.6. Expired carbon monoxide, plasma thiocyanate, blood nicotine, and cotinine concentrations in 248 habitual smokers of cigarettes according to Federal Trade Commission yield. Asterisks indicate significant difference compared with other yields.

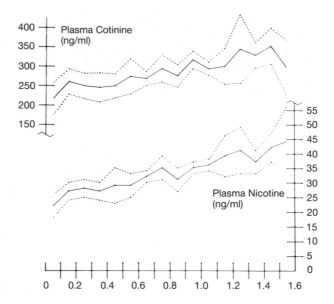

Fig. 10.7. Plasma cotinine and nicotine concentrations in habitual smokers according to Federal Trade Commission yield. Solid line indicates mean, dashed line 95 per cent confidence intervals (from Gori and Lynch[26]).

be a small decline in nicotine (and probably tar) intake. Our data suggest that intake from the ultra-low-yield cigarettes is less than expected from the shallow slope of decline of cotinine concentration with decreasing nicotine yield, with the result of a 30–50 per cent reduction in exposure to nicotine and tar. Thus, it appears reasonable to encourage smokers who are unable to quit to switch to ultra-low-yield cigarettes. However, presumably because actual nicotine yields are substantially reduced, ultra-low-yield cigarettes are perceived as unsatisfying, and the sales of these cigarettes in the population as a whole has remained and probably will remain low.

Spontaneous brand switching

Population sampling at one point in time represents a static observation of what is often a dynamic process. It does not address the natural history of smoking; that is, whether current smokers of low-yield cigarettes had previously smoked high-yield cigarettes. If so, did their intake of nicotine and other components of tobacco smoke change when they switched brands? This type of observation may be most relevant to developing policies about encouraging manufacturing of and switching to lower yield cigarettes.

Our laboratory participated in two studies in which a large population of smokers were randomly sampled in various locations in the United States.[26,28] Information about cigarette consumption, cigarette brands, and measurements of plasma cotinine and expired air carbon monoxide concentration was obtained. Six years later, subjects were contacted again to determine if they had changed brands.[28] Of an original group of 807 subjects, 197 were recruited for further study. Of these, 104 were smoking cigarettes of the same or similar smoking machine-determined yield as before. Sixty-two had switched to lower yield (0.2 mg or more reduction in nicotine delivery) and 31 switched to higher yield (0.2 mg or more increase in nicotine delivery; Fig. 10.8). Plasma samples and expired carbon monoxide were measured at approximately the same time of day at the baseline and on retesting.

Smokers who did not change yield showed a slight decrease in numbers of cigarettes smoked per day, but no change in cotinine or CO levels (Figs 10.7–10.9). Smokers who switched to lower-yield cigarettes had initially smoked cigarettes with higher nicotine yields (average 1.09 mg) and decreased an average of 38 per cent to a yield of 0.68 mg. Brand switching was associated with a reduction in cotinine and expired CO of about 20 per cent. However, these smokers also decreased their cigarette consumption by about 20 per cent (Fig. 10.8). Analysis of cotinine or CO per cigarette (Figs 10.9 and 10.10)

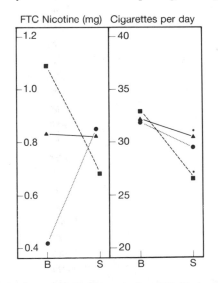

Fig. 10.8. Spontaneous brand switching study: US Federal Trade Commission nicotine yields and number of cigarettes smoked per day in groups at baseline (B) and at follow-up study (F). Symbols: ▲ = controls (*N* = 109), ■ = decreasers (*N* = 62), ● = increasers (*N* = 32). Asterisks indicate significant change from baseline to follow-up study.

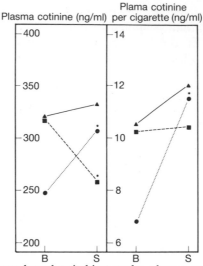

Fig. 10.9. Spontaneous brand switching study: plasma concentrations of cotinine and cotinine concentration normalized for cigarettes smoked per day. Symbols are as described in Fig. 10.8 (from Lynch and Benowitz[28]).

showed no change, despite reduction in yield. Thus, the smokers obtained the same dose of nicotine and CO from each cigarette even though the yield was lower. This observation is consistent with findings discussed previously showing that when switching from high- to low-yield (or in this case medium) cigarettes, full compensation is easily achieved. Reduction in daily exposure to tobacco smoke occurred primarily because the population who switched to lower-yield cigarettes smoked fewer cigarettes. Possibly, switching was part of an attempt of the individuals to reduce the health risks by smoking both lower yields and fewer cigarettes per day.

Switchers to high-yield cigarettes had smoked a lower-yield cigarette at the initial study (0.42 mg nicotine) and increased 102 per cent, to an average of 0.85 mg. They had significantly lower initial plasma cotinine. After switching, cotinine levels increased by 23 per cent and expired CO by 5 per cent (Figs 10.8 and 10.9). In this case, smokers did consume more nicotine and carbon monoxide per cigarette, although much less than predicted by the relative increase in machine yield. Since these subjects were smoking lower yield cigarettes and had lower cotinine levels than other groups before switching, it is possible that this group was composed of smokers in an early phase of developing tobacco dependence. This idea is supported by the observation that, after switching, cotinine levels rose to levels similar to those of the two other groups at baseline. Another study has also found increased nicotine and CO exposure after experimental switching to higher-yield cigarettes.[29]

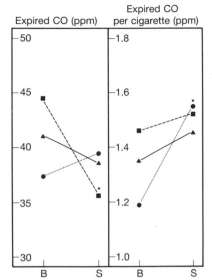

Fig. 10.10. Spontaneous brand switching study: expired air carbon monoxide (CO) concentration and CO concentration normalized for cigarettes smoked per day. Symbols and methods of data analysis are as described in Fig. 10.8. (From Lynch and Benowitz.[28])

10.4.   Summary and conclusions

It is desirable that people who cannot stop smoking reduce their exposure to harmful constituents of tobacco smoke. Due to the tendency to regulate intake of nicotine, reduction in exposure to tobacco smoke toxins does not automatically result from smoking cigarettes with lower smoking machine-determined yields. Evaluation of the potential health benefits of brand switching requires methods for estimating doses of toxins taken in by the smoker. Methods for estimating tar, nicotine, and carbon monoxide exposure have been reviewed. Due to individual differences in metabolism, markers of exposures are imperfect measures of dose. However, when used for within-subject comparisons or in large population samples, useful dosimetric information can be obtained.

We and other laboratories have studied the influence of cigarette yield on toxic exposures after experimental brand switching and in population studies, comparing self-selected brands. In addition, we have studied the consequences of spontaneous brand switching. Our data indicate that exposure to tobacco smoke toxins is reduced by only a small degree by switching from modern high- or middle- to low-yield cigarettes. A somewhat larger, but still small degree of reduction is achieved by switching to ultra-low-yield cigarettes, but most smokers find the cigarettes to be unsatisfying. Greater reduction in exposure to toxic chemicals may be accomplished if smokers are encouraged to smoke fewer (or at least not to smoke more) lower-yield cigarettes and to avoid smoking more of each cigarette. Our data indicate that switching to higher-yield cigarettes may result in greater exposure to tobacco toxins and escalation of yield should be strongly discouraged.

Acknowledgements

This research was supported in part by grants CA32389, DA02277, and DA01696 from the National Institutes of Health.

References

1. Wulf, H. C., Husum, B., and Niebuhr, E. (1983). Sister chromatid exchanges in smokers of high-tar cigarettes, low-tar cigarettes, cheroots and pipe tobacco. *Hereditas*, **98**, 225–8.
2. Yamasaki, E. and Ames, B. N. (1977). Concentration of mutagens from urine by absorption with the nonpolar resin XAD-2: Cigarette smokers have mutagenic urine. *Proceedings of the National Academy of Science*, **74**, 3555–9.
3. Florin, I., *et al.* (1980). Screening of tobacco smoke constituents for mutagenicity using the Ames' test. *Toxicology*, **18**, 219–32.
4. Jaffe, R. L., Nicholson, W. J., and Garro, A. J. (1983). Urinary mutagen levels in smokers. *Cancer Letters*, **20**, 37–42.
5. Sorsa, M., *et al.* (1984). Detection of exposure to mutagenic compounds in low-tar and medium-tar cigarette smokers. *Environmental Research*, **33**, 312–21.
6. Van Doorn, R., *et al.* (1979). Thioether concentration and mutagenicity of urine from cigarette smokers. *International Archives of Occupational and Environmental Health*, **43**, 159–66.
7. Young, J. C., Robinson, J. C., and Rickert, W. S. (1981). How good are the numbers for cigarette tar at predicting deliveries of carbon monoxide, hydrogen cyanide, and acrolein? *Journal of Toxicology and Environmental Health*, **7**, 801–8.
8. Russell, M. A. H. *et al.* (1986). Reduction of tar, nicotine and carbon monoxide intake in low tar smokers. *Journal of Epidemiology and Community Health*, **40**, 80–5.
9. Benowitz, N. L., Jacob, P., III, Jones, R. T., and Rosenberg, J. (1982). Interindividual variability in the metabolism and cardiovascular effects of nicotine in man. *Journal of Pharmacology and Experimental Therapeutics*, **221**, 368–72.
10. Benowitz, N. L. and Jacob, P., III (1984). Daily intake of nicotine during cigarette smoking. *Clinical Pharmacology and Therapeutics*, **35**, 499–504.
11. Benowitz, N. L., Kuyt, F., Jacob, P., III, Jones, R. T., and Osman, A-L. (1983). Cotinine disposition and effects. *Clinical Pharmacology and Therapeutics*, **34**, 604–11.
12. Benowitz, N. L., Jacob, P., III, Yu, L., Talcott, R., Hall, S., and Jones, R. T.

(1986). Reduced tar, nicotine, and carbon monoxide exposure while smoking ultra low- but not low-yield cigarettes. *Journal of the American Medical Association*, **256**, 241–6.

13. Benowitz, N. L. and Jacob, P., III (1984). Nicotine and carbon monoxide intake from high- and low-yield cigarettes. *Clinical Pharmacology and Therapeutics*, **36**, 265–70.

14. Benowitz, N. L., *et al.* (1983). Smokers of low-yield cigarettes do not consume less nicotine. *New England Journal of Medicine*, **309**, 139–42.

15. Kozlowski, L. T., *et al.* (1982). Estimating the yield to smokers of tar, nicotine, and carbon monoxide from the 'lowest yield' ventilated filter-cigarettes. *British Journal of Addiction*, **77**, 159–65.

16. Lombardo, T., Davis, C. J., and Prue, D. M. (1983). When low tar cigarettes yield high tar: Cigarette filter ventilation hole blocking and its detection. *Addictive Behaviours*, **8**, 67–9.

17. Schlotzhauer, W. S. and Chortyk, O. T. (1983). Effects of varied smoking machine parameters on deliveries of total particulate matter and selected smoke constituents from an ultra low-tar cigarette. *Journal of Analytical Toxicology*, **7**, 92–5.

18. Hoffmann, D., Adams, J. D., and Haley, N. J. (1983). Reported cigarette smoke values: A closer look. *American Journal of Public Health*, **73**, 1050–3.

19. West, R. J., *et al.* (1984). Does switching to an ultra-low nicotine cigarette induce nicotine withdrawal effects? *Psychopharmacology*, **84**, 120–3.

20. Gori, G. B., and Lynch, C. J. (1983). Smoker intake from cigarettes in the 1-mg Federal Trade Commission tar class. *Regulatory Toxicology and Pharmacology*, **3**, 110–20.

21. Gori, G. B. and Lynch, C. J. (1978). Toward less hazardous cigarettes. *Journal of the American Medical Association*, **240**, 1255–9.

22. Rickert, W. S., *et al.* (1983). A comparison of the yields of tar, nicotine, and carbon monoxide of 36 brands of Canadian cigarettes tested under three conditions. *Preventive Medicine*, **12**, 682–94.

23. Jaffe, J. H., Kanzler, M., and Friedman, L. (1981). Carbon monoxide and thiocyanate levels in low tar/nicotine smokers. *Addictive Behaviours*, **6**, 337–43.

24. Russell, M. A. H., Sutton, S. R., Iyer, R., Feyerabend, C., and Vesey, C. J. (1982). Long-term switching to low-tar low-nicotine cigarettes. *Journal of Addiction*, **77**, 145–58.

25. Young, J. C., Robinson, J. C., and Rickert, W. S. (1984). How good are the numbers for cigarette tar at predicting deliveries of carbon monoxide, hydrogen cyanide, and acrolein? *Journal of Toxicology and Environmental Health*, **7**, 801–8.

26. Gori, G. C. and Lynch, C. J. (1985). Analytical cigarette yields as predictors of smoke bioavailability. *Regulatory Toxicology and Pharmacology*, **5**, 314–26.

27. Hill, P., Haley, N. J., and Wynder, E. L. (1983). Cigarette smoking: Carboxyhaemoglobin, plasma nicotine, cotinine and thiocyanate versus self-reported smoking data and cardiovascular disease. *Journal of Chronic Diseases*, **36**, 439–49.

28. Lynch, C. J. and Benowitz, N. L. (1987). Spontaneous cigarette brand switching: Consequences for nicotine and carbon monoxide exposure. *American Journal of Public Health*, **77**, 1191–4.

29. Sepkovic, D. W., *et al.* (1984). Cigarette smoking as a risk for cardiovascular disease. V: Biochemical parameters with increased and decreased nicotine content cigarettes. *Addictive Behaviours*, **9**, 255–63.
30. Federal Trade Commission (1983). *Report of 'Tar', Nicotine and Carbon Monoxide of the Smoke of 208 Varieties of Cigarettes*. United States Federal Trade Commission.

11

Nicotine and the self-regulation of smoke intake

M. A. H. RUSSELL

Abstract

That smokers regulate their intake of smoke is not disputed and evidence is presented that nicotine is a major controlling factor, independent of other smoke components. The degree of regulation varies, but smokers tend to avoid exceeding their usual levels of blood nicotine. The average nicotine intake (heavy smokers) is about 1 mg per cigarette and with the known mechanical problems of compensatory increases in the intensity of smoking, nicotine intake is more difficult to maintain as the nicotine yield falls below about 0.4 mg per cigarette. Arguments are presented for maintaining cigarette nicotine yields at 1.0–1.3 mg, while continuing to lower the yields of tar, carbon monoxide, and other harmful gases.

11.1. Introduction

For many years, developed countries have had some form of national policy to reduce the harmfulness of cigarettes. In general, these policies have sought to bring about an overall reduction in the yields of cigarettes and have focused particularly on lowering the tar and nicotine yields. However, the limitations of this approach are becoming increasingly recognized and the purpose of this symposium was to review the situation and to consider whether new approaches should be adopted.

Two major problems confront programmes aimed at lowering both tar and nicotine yields of cigarettes. The first is the lack of acceptability of low yield brands. After their introduction to the UK in 1972, the market share of low-tar cigarettes (<10.5 mg) increased initially, but then flattened and has stuck obstinately at about 15–17 per cent.[1,2] Likewise, the proportion of smokers who regularly smoke a low-tar brand increased steadily until 1977, since when it too has fluctuated narrowly between 15 and 17 per cent. This resistance of the smoking population to using low-tar brands appears to have been largely unaffected by encouragement from health authorities for continuing smokers to switch to them, by the increased number and wider choice of low-tar brands available, and by the efforts of the industry to promote them. Thus, the data on market share and prevalence of use both tell the same story. Current low-tar cigarettes are insufficiently acceptable and only a minority of smokers will smoke them.

The second problem faced by current low-tar programmes is the tendency for smokers to regulate their smoke intake.[3-6] On switching to higher yielding cigarettes, they puff less hard and less frequently, inhale less deeply, and may smoke fewer cigarettes. Conversely, on switching to lower yield cigarettes, they smoke more intensively, inhale more deeply, and thereby undermine much of the potential benefit to their health.

What is the solution? Is there a case for medium nicotine low-tar cigarettes? The industry could certainly provide them. The answer is dependent on two main issues: the role of nicotine in the self-regulation of smoke intake, and the harmfulness of nicotine *per se* as opposed to other smoke products. The logic is quite simple.[7,8] If people smoke cigarettes to get nicotine, but die from the tar, it would be logical to reduce deliveries of tar and other harmful products not essential to acceptibility, but to maintain deliveries of nicotine at levels sufficient to satisfy smokers and to prevent their need to smoke more intensively to compensate for a drop in nicotine. One possible solution, therefore, is a gradual move towards a medium nicotine low-tar cigarette with the emphasis on low tar/nicotine yield ratios rather than low absolute yields. There is no reason why yields of carbon monoxide, nitrogen oxides, and other harmful components should not also be reduced.

As mentioned above, one concern about a medium nicotine low-tar approach is the potential harmfulness of nicotine itself. It undoubtedly has a role in smoking-related cardiovascular disease and may even make a small contribution to smoking-related cancers. These issues are covered elsewhere in this book. However, the approach is not logically negated by having to continue to accept those risks attributable to nicotine, unless there is any question about the value of reducing the intake of tar and other components. Nevertheless, the potential risks of nicotine raise legitimate concern that a medium nicotine low-tar approach should not tend to raise the intake of nicotine. The core issue, therefore, is the extent to which the acceptability of cigarettes and the self-regulation of smoke intake is determined by nicotine as opposed to all the other factors. My colleague, Martin Jarvis, has addressed the role of nicotine as the primary motive for continued smoking (chapter 12). It is my brief here to consider its role in the self-regulation of smoke intake. I shall also consider a further approach to less harmful cigarettes which emerges from an attempt to reconcile the nicotine intake data with the epidemiological data. Taken together, the findings of these two methodologies suggest that our programmes should not remain obsessed with simply lowering tar, but focus some attention on its quality and specific carcinogenicity as well as the quantity.

Finally, there is an extensive literature and several reviews on the self-regulation of smoke intake.[3-6] I reviewed this recently[6] and will focus here on those aspects most relevant to the concerns of this symposium.

11.2. Nicotine as a determinant of smoke intake

Tar and nicotine yields of commercial cigarettes are highly correlated. To determine the role of nicotine in the regulation of smoke intake requires control of all possible confounding factors (tar yield, draw resistance, flavour, burning characteristics, gas phase components, etc.). This is difficult to achieve. Various approaches have been tried, including studies of smoking behaviour in response to:

(1) variations in nicotine yield of cigarettes;

(2) nicotine administration by injection, infusion, ingestion, or via nicotine chewing gum;

(3) pharmacological blockade of nicotine actions;

(4) alteration of nicotine excretion by inducing changes in urinary pH.

Nicotine yield and smoking behaviour
It is obviously difficult to manipulate the nicotine yield of cigarettes without at the same time affecting other confounding factors. This has seldom been achieved, but was done more than 20 years ago by Finnegan *et al.*[9] in a study for which the experimental cigarettes were prepared by using a tobacco naturally low in nicotine and then adding nicotine to produce the higher yielding cigarette. The two cigarettes therefore differed only in their nicotine yield. This study was the first demonstration of the up-regulation of nicotine intake, although the authors seem to have been unaware of this aspect. Their purpose was to look for the development of withdrawal symptoms after 'blind' substitution of low nicotine cigarettes. This they found; but, in addition, as is clear in Table 11.1, about half the subjects increased their cigarette consumption on the low nicotine cigarette and had few withdrawal symptoms. Those subjects who did not up-regulate (compensate) by smoking more cigarettes, suffered more severe withdrawal symptoms which persisted throughout the month that they were smoking the low nicotine cigarettes. Although the changes in cigarette consumption were small, this is a crude measure of nicotine intake, and compensatory changes in puffing and inhalation might have occurred also. One or two other studies have attempted to examine the effects of changes in nicotine yield independently of other factors (although with only limited success) and the results have suggested that the smoking pattern is affected more by the delivery of nicotine than that of tar.[10,11]

Table 11.1. Effect of nicotine yield on cigarette consumption

Subject Group	Average daily cigarette consumption per group			
	Usual brand	1.96 mg nicotine	0.34 mg nicotine	1.96 mg nicotine
No withdrawal symptoms ($n = 6$)	26.9	26.6	30.9	26.8
Transient mild withdrawal symptoms ($n = 6$)	22.4	22.0	26.5	23.9
Transient moderate withdrawal symptoms ($n = 3$)	23.6	28.3	28.6	27.6
Severe withdrawal symptoms persisting for 1 month ($n = 9$)	25.0	24.7	24.6	24.9

Abstracted from Finnegan et al.[9] The nicotine yield of the subjects' usual brand was not given, but an average yield for cigarettes of that period would have been around 2.0 mg. The experimental cigarettes were made from tobacco naturally low in nicotine (i.e. 0.34 mg yield), but nicotine was added to this tobacco to provide the cigarettes with the higher yield (1.96 mg). The order of administration was as shown from left to right. The exact period on each brand was not clearly stated but was about 4 weeks on the usual and low nicotine brands, and 2 weeks for each period on the 1.96 mg brand. Withdrawal symptoms on switching to the low nicotine cigarettes were graded as follows: mild = vague lack of satisfaction initially; moderate = missed their usual brand but eventually adapted to the low nicotine cigarettes; severe = missed their usual brand even after 1 month and experienced irritability, poor concentration and/or hunger-like feelings.

Nicotine administration

In 1942, Johnston[12] gave smoking doses of nicotine by hypodermic and intravenous injection to 35 volunteers. He reported that non-smokers found it unpleasant, but 'smokers almost invariably thought the sensation pleasant and, given an adequate dose, were disinclined to smoke for some time thereafter'. The nicotine injections also relieved the withdrawal cravings of smokers deprived of tobacco. However, this pioneering study was neither 'blind' nor controlled. Twenty-five years later, Lucchesi et al.[13] showed, under double-blind conditions, that slow intravenous infusion of nicotine at a rate of 4 mg per hour had an inhibitory effect on *ad libitum* smoking compared with saline control. There was a 27 per cent reduction in the number of cigarettes smoked per session, and subjects also took significantly fewer puffs and discarded their cigarettes earlier. Although the amount of suppression was small in relation to the dose of nicotine given, the slow infusion would not have simulated the intermittent bolus dosage from smoking and any decrease in inhalation would have been undetected. The Lucchesi et al. findings were not confirmed by a subsequent study[14] which failed to show an inhibitory effect on puffing during and after intravenous nicotine. One explanation is that the subjects in the second study were so hampered by apparatus for physiological measures that they were unable to hold their cigarettes and smoke naturally. Smoking involved puffing from a holder, which was positioned obtrusively close to the subject's face and contained a continuously lighted cigarette, replaced as necessary by the experimenter. This experimental situation and its demand characteristics may have swamped the factors which normally control smoking. In a recent intravenous nicotine preloading study in our laboratory, which is not yet published, the findings support those of Lucchesi *et al.* in showing a clear inhibitory effect on subsequent smoking behaviour. Furthermore, the down-regulation of nicotine intake was so precise that, despite preloading to an average blood nicotine concentration in excess of 25 ng/ml, there was no tendency for the subjects to overshoot their normal blood nicotine levels when *ad libitum* smoking was permitted.

Studies using other forms of nicotine administration show similar results. One early study showed that ingestion of 10 mg nicotine daily had a small, but significant inhibitory effect, and reduced daily cigarette consumption by 8 per cent.[15] Bearing in mind that the average nicotine intake from smoking is 20–30 mg/day and that the systemic bioavailability of oral nicotine is not more than 30 per cent, the small size of the effect is understandable. Nicotine chewing gum also has a small, but statistically significant inhibitory effect on *ad libitum* smoking compared with placebo. Fewer cigarettes are smoked after chewing it[16] and inhalation is reduced as measured by a drop in carboxyhaemoglobin levels.[17] Further evidence of the precision of down-regulation of nicotine from smoking is provided by an early study of gum use.[17] A group of 43 smokers were instructed to continue to smoke as they wished, but to also take ten pieces of 2 mg nicotine gum per day at a rate of about one per hour (this pattern of gum use generally produces steady-state blood nicotine levels around 10 ng/ml). Despite the extra nicotine from the gum, blood nicotine levels were not increased, indicating that smokers had down-regulated their intake from smoking. The peak blood nicotine just after a cigarette averaged 27.4 ng/ml when they were taking gum and smoking *ad libitum*, compared with the similar level of 30.1 ng/ml when smoking normally without the gum.

In summary, these studies involving the administration of nicotine provide a specific test of the effect of nicotine on the regulation of smoke intake without confounding from other factors. Taken together they provide sound evidence that nicotine is an important factor in down-regulation. However, by their nature they are limited to testing down-regulation and cannot inform about the up-regulation of smoke intake. It has been suggested that the precision of down-regulation may be attributable to avoidance of the aversive toxic effects of excessive doses.[18] Whatever the mechanism, the sensitivity of the avoidance of excessive intake, the precision of down-regulation, and the lack of any tendency to overshoot to higher blood nicotine levels should be reassuring to those who fear that a medium nicotine low-tar cigarette might lead to unacceptable increases in nicotine intake.

Nicotinic blockade

A novel approach was the use of a centrally acting nicotine antagonist, mecamylamine, which increased the number of puffs and the number of cigarettes smoked by 25 and 30 per cent respectively, compared with a placebo.[19] In contrast, pentolinium, an antagonist with predominantly peripheral actions, had no effect on smoking behaviour. This study confirms that smoking behaviour is modified to regulate nicotine intake and that this effect is mediated by its actions in the brain. Its actions at peripheral sites appear not to be implicated.

Urinary nicotine excretion rate

Urinary pH can be manipulated by taking ammonium chloride which makes it more acidic, or sodium bicarbonate which makes it alkaline. Under acid conditions, up to 15 per cent or more of nicotine is excreted unchanged in the urine (the remainder being metabolized). However, under alkaline conditions urinary nicotine excretion is negligible. Schachter showed that when subjects received tablets to acidify their urine and increase nicotine excretion their cigarette consumption increased. Conversely, when nicotine excretion was reduced by raising urinary pH, they smoked fewer cigarettes.[20] This study provides clear evidence that smoking behaviour is modified to regulate nicotine intake independently of other factors. In a recent study, Benowitz and Jacob[21] manipulated urinary pH and confirmed Schachter's findings. They also measured blood nicotine concentrations and showed that they were not increased when urinary nicotine excretion was reduced by sodium bicarbonate (i.e. perfect down-regulation), but were 15 per cent lower during urinary acidification with ammonium chloride (i.e. incomplete up-regulation).

These four classes of study employing a variety of approaches, all designed to test specifically the role of nicotine, are remarkably concordant in their findings. They show that nicotine intake is down-regulated with great precision to avoid exceeding the blood nicotine concentrations of usual smoking. On the other hand, up-regulation (or compensation) appears to be less sensitive and is partial or incomplete. However, blood nicotine concentrations were measured in only two of the studies so that data on the quantitative aspects of intake regulation are rather limited. This is further considered with the findings of brand-switching studies in which tar yield and other confounding factors are not all controlled.

11.3. Brand-switching studies

Apart from the few exceptions discussed in the preceding section, studies of the effect on smoking behaviour of switching to higher or lower yield cigarettes are limited by lack of control for tar yield and most of the other confounding factors. Nevertheless, they have value with respect to other relevant questions. Out of more than 100 studies of this kind, attention is focused here on some of the few that include measures of blood nicotine concentrations. Data from four such studies[22-25] are summarized in Table 11.2. In two cases the switch was short-term, 5 hours and 24 hours. Two involved switching to a higher as well as lower yield brand. All involved switching to a lower yield brand, two to 'low', two to 'ultra-low'. Collectively, they impart much relevant information.

It can be seen in Table 11.2 that there is no tendency for blood nicotine concentrations to increase after short-term switching to high yield brands. This confirms the picture presented in the preceding section of the precision with which nicotine intake is down-regulated to avoid excessive blood levels. Whether this would continue to hold after long-term switching is not known. The lower sensitivity of up-regulation is also confirmed. In all cases blood nicotine concentrations were lower after switching to lower yield brands, although the drop was less than would have occurred without a change to more intensive smoking. In no cases did the number of cigarettes smoked increase significantly after switching to lower yield cigarettes, indicating that the compensatory changes were confined to altering the way that individual cigarettes were smoked. In the case of the ultra-low brands, complete up-regulation would have been impossible without increasing the number of cigarettes smoked. On the low tar brands, however, complete up-regulation of nicotine intake would have been relatively easy in mechanical terms. Why this was not accomplished is not known, but it suggests that smokers may tolerate a modest drop in nicotine intake more readily than they do an increase above their usual smoking levels. It is noteworthy that the degree of up-regulation appears to become established quickly and then remains constant, in that blood nicotine concentrations at 10 days and 10 weeks after switching were similar to those at 1–2 days (Table 11.2).

11.4. Acceptability and intake regulation

Having shown that the regulation of smoke intake is controlled by nicotine, it is relevant to consider here how its role in this may be linked with the acceptability of lower yield cigarettes, although this aspect is covered comprehensively by Martin Jarvis (chapter 16). Besides nicotine intake,

Table 11.2. Changes in plasma nicotine concentrations after switching to high, low or ultra-low yield cigarettes (values shown are group means)

	Brand of cigarette	Tar yield (mg/cig)	Nicotine yield (mg/cig)	Plasma nicotine (ng/ml)
Russell et al. (1975)				
(n = 10)	Usual	18.0	1.34	30.1
	Ultra-low tar	1.0	0.14	8.5
	High tar	32.0	3.2	29.2
% reduction (low v. usual)		94	90	72
Ashton et al. (1979)				
(n = 12)	Usual	18.0	1.4	39.6
	Low tar	6.5	0.6	30.9
	High tar	26.5	1.84	42.8
% reduction (low v. usual)		64	57	22
Russell et al. (1982)				
(n = 12)	Usual	17.4	1.33	32.4
	Middle tar	19.0	1.3	35.2
	Usual	17.4	1.33	33.2
	Low tar (2 days)	10.9	0.7	20.8
	Low tar (8 wks)	10.9	0.7	22.7
	Low tar (10 wks)	10.9	0.7	22.8
% reduction (low v. usual)		39	47	33
West et al. (1984)				
(n = 11)	Usual	14.2	1.3	22.0
	Ultra-low (1 day)	1.0	0.1	11.2
	Ultra-low (3 days)	1.0	0.1	7.3
	Ultra-low (10 days)	1.0	0.1	10.2
% reduction		93	92	57

Abstracted from Russell *et al.* (1975), Ashton *et al.* (1979), Russell *et al.* (1982), and West *et al.* (1984).[22–25] On switching to lower yield cigarettes the blood nicotine concentrations are significantly less than on the usual brands. However, the reductions in blood levels are proportionately less than the reductions in nicotine yields indicating that some compensatory increase in puffing and inhalation occurred although it was not sufficient to compensate completely for the drop in yield (i.e. partial up-regulation). Cigarette consumption did not change significantly in any of the studies, except for the switch to the high tar cigarette in the Russell *et al.* (1975) study. On switching to a high yield brand or another middle-tar brand the plasma nicotine levels did not change significantly. The drop at 8 and 10 weeks after switching to a low-yield brand was similar to that at 2 days after switching (Russell *et al.* 1982). The plasma nicotine concentrations in the West *et al.* study were 'trough' levels, whereas the other three studies used 'peak' levels 2 minutes after a cigarette.

two other factors are important in determining the satisfaction and acceptability of cigarettes: namely, the flavour and the mechanical factors involved with intake regulation. With down-regulation, for example, no mechanical problems arise and there is clearly little difficulty about taking smaller puffs and inhaling less deeply after switching to higher yield cigarettes. Switching to lower yield cigarettes, on the other hand, may present a number of difficulties. Puff volume and inhalation can be increased to maintain intake and nicotine concentrations, but only up to a point. After this, compensatory changes in smoking pattern become progressively more intrusive, then aversive, and eventually, with very low yields, complete compensation becomes impossible. This is affected profoundly by the hyperbolic nature of the mathematical relationship between the nicotine concentration in the smoke and the volume of smoke required to deliver a given dose.[26]

To illustrate the problem, let the starting point be a middle tar cigarette which delivers 1.2 mg nicotine in 400 ml of mainstream smoke from a machine set to take ten 40-ml puffs. Next let us consider a series of lower yield cigarettes having machine-smoked yields, derived from the same machine setting of ten 40-ml puffs, as follows: 0.8, 0.6, 0.4, and 0.2 mg nicotine. To obtain 1.2 mg nicotine from these lower yield cigarettes, the smoker would have to inhale 600, 800, 1200, and 2400 ml of smoke per cigarette, respectively. It is clear that things soon get difficult and then impossible as the smoker reaches the steep part of the curve.

The issue is further complicated by various characteristics of the particular brand of lower yield cigarette and of the individual smoker. Low yield cigarettes vary greatly in the extent to which they can be 'oversmoked' to exceed their standard machine-smoked yields. The factors involved include the length of tip overwrap and the extent to which yield reductions are achieved by filtration, ventilation, or lowering the quantity of tobacco or its nicotine content. On average, heavy smokers take in about 1 mg nicotine per cigarette, but there is wide variation between individual smokers[27,28] and individual dosage requirement would determine at what point on the nicotine yield 'league table' a given smoker would hit the steep part of the compensation curve described above. With UK cigarettes, the majority of smokers appear to first encounter these mechanical problems when nicotine yields fall below about 0.6 mg, and cigarettes with yields below 0.4 mg nicotine are currently almost unsaleable. The longer tip overwraps in US low yield brands make it easier for smokers to exceed nominal yields which have been artifactually lowered.[29] A final point regarding individual differences between smokers is the extent to which they smoke to obtain a given nicotine boost per cigarette as opposed to a particular steady-state plateau level. In the latter case up-regulation would be easier in that it could be assisted by smoking more cigarettes – a strategy not open to the former.

The change in flavour can affect smokers' reactions to a different brand irrespective of its nicotine delivery. Among the studies shown in Table 11.2, one included a switch to a higher yield brand[22] and another to a brand with a yield similar to that of the usual brand.[24] In both cases, the smokers reported decreases in satisfaction, although their blood nicotine levels were unchanged. These two examples were short-term switches and I am not aware of data on how long it takes for smokers to adjust to a change in flavour which is not accompanied by a drop in blood nicotine level or any need for undue raising of the intensity of smoking to maintain it. In my view, flavour is a secondary factor so that any loss of satisfaction attributable to a change in flavour would be short-lived, so long as nicotine intake was sufficient to provide reinforcement of the new flavour. A useful lesson can perhaps be learnt from recent history. Between 1955 and 1975 the UK smoking population achieved a virtually complete switch from high tar plain unfiltered cigarettes to middle tar filter-tipped brands. This switch involved adjustment to considerable changes in flavour, but nicotine intake would have been relatively unimpaired. The ease with which it was accomplished points to the transience and fundamental unimportance of the flavour factor.

In keeping with the lower sensitivity and incompleteness of up-regulation of nicotine intake, smokers seem to tolerate a drop in nicotine intake to roughly two thirds of their usual levels with some initial loss of satisfaction, but they adapt gradually to the lower level over 2–3 weeks.[24] In the longer term study of switching to a low tar cigarette, shown in Table 11.2, apart from being too weak, the initial loss of satisfaction on the low tar brand was no greater than on switching to another middle tar brand. By 8–10 weeks, the low tar cigarette was preferred to the usual brand and all twelve subjects thought that they could change to it permanently, despite still showing blood nicotine and cotinine levels down to two-thirds of their previous levels. The contrast of these results with the low acceptability of lower yield brands among smokers in the general population suggests that many more might switch if they could be persuaded to tolerate the initial loss of satisfaction for long enough to adjust and get used to lower yield smoking. However, for reasons outlined above, as nicotine yields fall below 0.6 mg per cigarette up-regulation becomes mechanically difficult and blood nicotine levels cannot be maintained even at an acceptable two-thirds of usual levels. On switching to ultra-low yield brands, blood nicotine drops to about one-third of the usual smoking levels (Table 11.2), some but not all of the nicotine-related withdrawal symptoms are experienced and the effects of the drop in nicotine intake are compounded by loss of flavour and the aversiveness of high intensity smoking so that satisfaction ratings drop to zero or slight.[25]

11.5. Nicotine intake from self-selected brands

Some support for the controlled brand-switching studies comes from several large-scale studies of nicotine intake from self-selected brands. In 1980, we reported a study of the blood nicotine levels of 330 regular smokers who had been smoking their usual brand in their usual way.[30] Some were smokers of low tar brands, others of middle or high tar brands. Most had been smoking their own particular brand for a year or more and, since low tar brands had been available for only a few years, the low tar smokers could be classified as long-term switchers. Blood was sampled during the afternoon or evening by which time steady-state conditions would have occurred in most cases. The most striking feature was the wide variation between individuals, from 4 to 72 ng/ml around a mean of 33 ng/ml. It was also evident that high blood nicotine levels of 40–60 ng/ml or more could be obtained from low tar cigarettes having nicotine yields of about 0.6 mg, whereas some smokers of high tar cigarettes with nicotine yields of 1.7 mg or more had blood nicotine levels below 20 ng/ml. Despite large differences in the machine-smoked yields of the cigarettes blood nicotine levels were only slightly lower in those on the lower yield brands. The correlation between blood nicotine and nicotine yield of cigarette, though statistically significant, was low (0.21, $P<0.001$) showing that the nicotine yield of the cigarettes accounted for only 4.4 per cent of the variation in blood nicotine levels.

These findings have been widely replicated and are also essentially the same in the US where the range of nicotine yields extends well below 0.6 mg, which was the lower limit in our samples of UK smokers.[31-34] It is clear from all these studies that the nicotine intake of smokers is determined more by the way they puff and inhale than by the nicotine yield of the cigarettes they smoke. It is also apparent that the differences in the blood nicotine concentrations of low yield and higher yield smokers are proportionally less than the differences between the machine-smoked yields of their brands (Table 11.3). This suggests that the lower yield smokers were smoking and inhaling more intensively, presumably to compensate for the lower yields. Indeed, the degree of up-regulation may have been greater than is apparent from the data in Table 11.3, in that the self-selected low-tar smokers may, before switching, have had lower blood nicotine concentrations than the other smokers.

From the practical viewpoint, the potential health benefits of lower yield cigarettes do not appear to be completely undermined by compensatory changes in the way they are smoked. As Table 11.3 shows, the smokers of low tar cigarettes had a lower intake of tar, nicotine, and carbon monoxide than the smokers of higher yielding brands. On average, their estimated intake of tar was 25 per cent lower, their intake of nicotine about 15 per

162 *M. A. H. Russell*

Table 11.3. Average percentage reductions in tar, nicotine, and carbon monoxide intake by low-tar smokers in comparison with smokers of higher yield brands

	Non-low-tar smokers ($n = 241$)	Low-tar smokers ($n = 151$)	% Reduction
Cigarette data			
Tar yield (mg/cig)	17.3	9.2	46.8
Nicotine yield (mg/cig)	1.42	0.86	39.3
CO yield (mg/cig)	16.6	11.0	34.1
T/N ratio	12.3	10.8	12.1***
T/CO ratio	1.06	0.85	19.8***
Intake measures			
Plasma nicotine (ng/ml)	38.3	31.8	17.0***
Plasma cotinine (ng/ml)	379	333	12.1*
COHb (%)	7.81	7.06	10.6**
Index of tar intake			
TI (Nic)	469	341	27.3***
TI (Cot)	4601	3543	23.0***
TI (CO)	8.13	5.93	27.1***

Note: The plasma cotinine data are based on smaller samples of 146 non-low-tar smokers and 95 low-tar smokers, but these subsamples did not differ from the remainder of the subjects in any of the other measures used. The percentage reduction in COHb was calculated after subtracting 0.7 to correct for the background level in non-smokers. The index of tar intake was derived from the measured blood level of a marker and the ratio of the tar to marker yields of the cigarette. For example, using nicotine as the marker, TI (Nic) = Plasma nicotine × T/N yield ratio. The cigarette yield ratios and indices of tar intake were computed for individual smokers before averaging them to obtain the group means. Statistical significance of differences are based on t tests between non-low-tar smokers: ***$P < 0.001$, **$P < 0.01$, *$P < 0.05$. From Russell *et al.*[34]

cent lower (17 and 12 per cent as measured by blood nicotine and cotinine, respectively), and their intake of carbon monoxide was about 10 per cent lower.[34] The lower blood nicotine concentrations are consistent with the drop in nicotine intake shown in the methodologically more rigorous experimental forced switching studies. Although the index of tar intake is an estimate and assumes a tar to nicotine yield ratio to the smoker equivalent to that of the machine-smoked yields, this approach worked in practice. For example, when the nicotine concentrations of the low tar smokers were predicted using carbon monoxide as a marker, the predicted mean level of 31.9 ng/ml was almost identical to the measured mean of 31.8 ng/ml. Published data on the effect of intensity of smoking on the tar to nicotine yield ratio have so far given conflicting results. This important question is covered in more detail elsewhere in this book (chapter 7). Finally, the fact that the 25 per cent reduction in estimated tar intake was more than the 15 per cent reduction in nicotine intake is attributable to the lower tar to nicotine yield ratios of the low tar brands. This illustrates the importance of focusing in the future on lowering the tar to nicotine yield ratios. One approach to this is a medium nicotine low tar cigarette.

11.6. Why are cigarettes getting less harmful?

Several epidemiological studies using classical case-control and prospective survey methods have shown that the risks of lung cancer among smokers of middle tar filter-tipped cigarettes are lower than those in smokers of high tar plain cigarettes without filters. Unfortunately, this evidence is not conclusive. One problem is that the brands were self-selected. Another is that brands used over decades of a smoking life may vary and precise data are difficult to obtain. However, more convincing evidence is provided by recent studies based on secular trends. These show a substantial decline in lung cancer mortality among younger men and women which is far greater than could be explained by the relatively small changes in cigarette consumption and the prevalence of smoking,[35] indicating that cigarettes have become less harmful and that continuing smokers are less likely to get lung cancer today than they were in the past. It should be emphasized that the comparison on which these findings are based is essentially between the effects of smoking predominantly plain high tar cigarettes before the mid-1950's and predominantly filter-tipped cigarettes (of the type now rated as middle tar) since the 1960's. It is too early for epidemiological methods to assess the cancer risks of the current type of low tar cigarettes with a slightly improved tar to nicotine ratio.

So, we can agree that middle tar filter-tipped cigarettes of the type used since the mid-1950's are less harmful than the older type of high tar plain cigarettes. What is the reason? This finding has been widely attributed to the reduction in tar yields. For example, on page 87 of its latest report, the Royal College of Physicians states that 'There is no obvious explanation for these falls in national lung cancer death rates other than the changes in the tar delivery of cigarettes'.[36] In my view, this interpretation is incorrect. First, there are other highly plausible explanations which we shall come to shortly. Secondly, acceptance of the RCP explanation would require rejection of the evidence for nicotine intake regulation (nicotine titration) which has been discussed above. The two models are shown in their extreme forms in the figure.

Data on the machine-smoked yields of cigarettes since the 1930's are available,[37] but no suitable data on the steady-state blood nicotine concentrations of smokers were gathered before the early 1970's. Between 1972 and

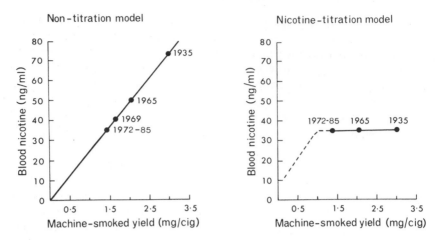

Fig. 11.1. Alternative models of the relationship between cigarette nicotine yields and average blood nicotine levels in smokers over the past 50 years and projected into the future. The non-titration model predicts higher nicotine levels in earlier years when average yields were higher. The nicotine-titration model postulates no change in nicotine levels over a wide range of yields. The dotted line suggests that with progressive reductions in yield nicotine titration will eventually break down.

1985, the sales-weighted nicotine yields of UK cigarettes did not decline as they had done in previous years (and as tar yields have continued to do), but remained static at about 1.3 mg per cigarette.[1,2] Likewise, over this period the steady-state peak blood nicotine concentrations of UK smokers have shown consistent average levels of around 30–35 ng/ml,[30,34] which are similar to those of US smokers.[31–33] In Fig. 11.1, the 1972–85 data on sales-weighted nicotine yields (1.3 mg) and peak steady-state blood nicotine concentrations (33 ng/ml) have been taken as the known starting point for the two models, and the blood nicotine levels for earlier years have been projected upwards or outwards along the axis for known machine-smoked yields for earlier years, on the assumption of no titration or perfect titration of nicotine intake, respectively.

The two models reflect the extremes of opposing views and the true situation is likely to have been somewhere between the two. However, the experimental evidence discussed above suggests that the situation must have been closer to that expressed by the nicotine titration model and that the non-titration model is widely off the mark. The precision with which smokers down-regulate their nicotine intake to avoid higher blood levels suggests that excessive intake may be aversive. However tobacco is used, the average blood nicotine levels are similar to or less than those from cigarette smoking. Thus, with smoking of pipes or cigars, tobacco chewing or use of 'wet' oral snuff or 'dry' nasal snuff, the average blood nicotine levels of regular users are almost invariably between 30–35 ng/ml. As with cigarette smoking, it would be relatively easy to obtain higher blood nicotine levels with other forms of tobacco use if it were rewarding. The similarity of the levels which users choose to obtain from such a wide variety of modes of use is unlikely to be coincidental. It would be stretching reason to assume that smokers of high tar, high nicotine cigarettes before the 1950's had average blood nicotine levels greatly in excess of about 40–45 ng/ml. Since tar and nicotine yields were highly correlated and the tar to nicotine yield ratios fairly constant until the recent introduction of low tar cigarettes, the tar intake of pre-1950 smokers is unlikely to have differed greatly from that of middle-tar smokers of today.

It has been mentioned that the average intake of systemically available nicotine is about 1 mg per cigarette in heavy smokers. This is the amount required to generate blood levels of 30–40 ng/ml, suggesting that higher yielding cigarettes delivering 1.5 mg nicotine or more are 'undersmoked'. The precision of nicotine intake regulation is reduced only when it switches from down-regulation to up-regulation at about the 1 mg per cigarette level. As has been discussed, the loss of precision is partly due to the tolerance of smokers to a modest drop in blood nicotine levels and partly due to the mechanical problems of up-regulating intake from lower yield cigarettes. Thus, the predicted curve for blood nicotine levels of smokers in the event of a decline in sales-weighted nicotine below 1.3 mg per cigarette (shown in the figure as the broken line in the titration model) shows the precision of intake regulation beginning to break down at yields below about 1 mg nicotine per cigarette.

If the argument above is correct, the change in the mid-1950's from high tar plain cigarettes to middle tar filter-tipped cigarettes is unlikely to have resulted in more than a modest reduction in tar intake to smokers lungs. The reduction would have been considerably less than could account for the major reductions in lung cancer mortality. What other factors could account for the epidemiological findings? One possible explanation is that the unventilated filters of middle tar cigarettes may have produced some qualitative change in the tar which lowered its 'specific carcinogenicity', over and above the effect of lowering overall tar yields. Another change, which occurred at about the same time as the introduction of unventilated filters, was the introduction of reconstituted tobacco sheets (RTS) in the manufacture of cigarettes. This is more economical and uses the whole of the tobacco plant, stalks, and all. Fortuitously, smoke from cigarettes made from RTS is less carcinogenic.[38] Thus, the same amount of tar taken in from cigarettes of the mid-1950's and before would be more likely to cause lung cancer than a similar dose from more modern brands (assuming that there is some validity to the animal models for testing carcinogenicity).

To summarize this section, it is evident that since the mid-1950's cigarettes have become less harmful with regard to their capacity to cause lung cancer. In my view, the evidence suggests that this is more likely to be due mainly to fortuitous reductions in the specific carcinogenicity of the tar they produce. The substantial reductions in tar yields of cigarettes are unlikely to have made more than a minor contribution to the reductions in lung cancer mortality observed so far, because it is unlikely that the switch to middle tar filter-tipped cigarettes was accompanied by appreciable reductions in tar intake into the lungs of smokers. It is too early for epidemiological studies to detect the effects on lung cancer mortality of low tar cigarettes introduced in the mid-1970s. However, studies based on intake measures, and the improved tar to nicotine yield ratios of these cigarettes, indicate that the tar intake of smokers who switch to them is substantially reduced. It follows that major reductions in lung cancer mortality can be predicted as more smokers use them and as the industry moves gradually closer to the production of medium nicotine low tar cigarettes.

11.7. Summary and conclusions

1. Smokers self-regulate their smoke intake and this is controlled by nicotine independently of other factors such as tar, carbon monoxide, and other gas phase components of the smoke.

2. Smokers down-regulate their nicotine intake with great precision to avoid exceeding their usual levels of blood nicotine, but up-regulation (or compensation) is less sensitive and is usually incomplete. This may be partly due to the aversiveness of intensive puffing and inhalation.

3. On balance, smokers seem to tolerate a drop in blood nicotine to roughly two-thirds of their usual levels with some initial loss of satisfaction, but they adapt gradually to the lower levels over two or three weeks. Greater reductions occur when fuller up-regulation is mechanically difficult or impossible (e.g. on ultra-low yield cigarettes) and generate increasing loss of satisfaction and some nicotine withdrawal effects.

4. The average nicotine intake of heavy smokers is about 1 mg per cigarette. In view of intake regulation (nicotine titration), and the precision of down-regulation in particular, it is unlikely that this was greatly exceeded when cigarettes delivered 2–3 mg nicotine per cigarette. Since the tar and nicotine yields were highly correlated, it follows that the switch from high tar plain cigarettes to middle tar filter-tipped cigarettes, which took place around the mid-1950s, cannot have been associated with any appreciable reductions in tar intake. It follows further that other explanations are required to account for the decline in the lung cancer mortality of UK smokers. It is suggested that an effect of

the filters and/or the use of reconstituted tobacco sheets have resulted in qualitative changes in the tar which have reduced its specific carcinogenicity.

5. In view of the lower sensitivity of up-regulation of nicotine intake and the mechanical problems of compensatory increases in the intensity of smoking, nicotine intake (and that of tar) declines progressively as nicotine yields fall below about 1 mg per cigarette. Likewise, acceptability and satisfaction decline steadily from about this point. Although smokers are able to adapt to lower intakes from cigarettes yielding about 0.6 mg nicotine, relatively few are motivated to persist long enough to achieve a long-term switch without specific support or encouragement.

6. To overcome the problems of lack of acceptability and compensatory up-regulation of nicotine intake, the low tar programme should adopt a strategy for a gradual move towards medium nicotine low tar cigarettes. Maintaining nicotine yields over the range of 1.0–1.3 mg is unlikely to result in any increase in nicotine intake which would add to those risks already attributable to nicotine. Any potential disadvantage in this respect would be far outweighed by the reductions in tar intake which could be achieved by getting the national sales-weighted tar to nicotine yield ratio down to 10 or less.

7. Besides focusing more on the tar to nicotine ratio of cigarette yields, advisory committees should place more emphasis on consideration of the qualitative characteristics and carcinogenicity of tar from different brands of cigarette and abandon the hitherto simplistic quantitative approach. Likewise, attention should be given to publication and lowering of yields of nitrogen oxides and other gas phase components which may be more harmful than carbon monoxide.

Acknowledgements

I thank the Medical Research Council for financial support over many years and Evelyn Langford for secretarial help.

References

1. Jarvis, M. J. and Russell, M. A. H. (1985). Tar and nicotine yields of UK cigarettes 1972–83: Sales-weighted estimates from non-industry sources. *British Journal of Addiction*, **80**, 429–34.
2. Jarvis, M. J. and Russell, M. A. H. (1986). Sales-weighted tar, nicotine and carbon monoxide yields of UK cigarettes: 1985. *British Journal of Addiction*, **81**, 579–81.
3. Russell, M. A. H. (1978). Self-regulation of nicotine intake by smokers. In *Behavioural effects of nicotine* (ed. K. Battig), pp. 108–22. Karger, Basel.

4. McMorrow, M. J. and Foxx, R. M. (1983). Nicotine's role in smoking: an analysis of nicotine regulation. *Psychological Bulletin*, **93**, 302–27.
5. Gritz, E. R. (1980). Smoking behaviour and tobacco abuse. In *Advances in substance abuse* (ed. N. K. Mello), pp. 91–158. JAI Press, Greenwich, Connecticut.
6. Russell, M. A. H. (1987). Nicotine intake and its regulation by smokers. In *Tobacco smoking and nicotine*, (ed. Martin, W. R., Van Loon, G. R., Iwamoto, E. T., and Davis, L.) pp. 25–50. Plenum, New York.
7. Russell, M. A. H., Wilson, C., Patel, U. A., Cole, P. V., and Feyerabend, C. (1973). Comparison of the effect on tobacco consumption and carbon monoxide absorption of changing to high and low nicotine cigarettes. *British Medical Journal*, **4**, 512–6.
8. Russell, M. A. H. (1976). Low-tar medium nicotine cigarettes: a new approach to safer smoking. *British Medical Journal*, **1**, 1430–3.
9. Finnegan, J. K., Larson, P. S., and Haag, H. B. (1945). The role of nicotine in the cigarette habit. *Science*, **102**, 94–6.
10. Goldfarb, T. L., Gritz, E. R., Jarvik, M. E., and Stolerman, I. P. (1976). Reactions to cigarettes as a function of nicotine and tar. *Clinical Pharmacology and Therapeutics*, **19**, 762–72.
11. Herning, R. I., Jones, R. T., Bachman, J., and Mines, A. H. (1981). Puff volume increases when low-nicotine cigarettes are smoked. *British Medical Journal*, **283**, 187–9.
12. Johnston, L. M. (1942). Tobacco smoking and nicotine. *Lancet*, **ii**, 742.
13. Lucchesi, B. R., Schuster, C. R., and Emley, G. S. (1967). The role of nicotine as a determinant of cigarette smoking frequency in man with observations of certain cardiovascular effects associated with the tobacco alkaloid. *Clinical Pharmacology and Therapeutics*, **8**, 789–96.
14. Kumar, R., Cooke, E. C., Lader, M. H., and Russell, M. A. H. (1977). Is nicotine important in tobacco smoking? *Clinical Pharmacology and Therapeutics*, **21**, 520–9.
15. Jarvik, M. E., Glick, S. D., and Nakamura, R. K. (1970). Inhibition of cigarette smoking by orally administered nicotine. *Clinical Pharmacology and Therapeutics*, **11**, 574–6.
16. Kozlowski, L. T., Jarvik, M. E., and Gritz, E. R. (1975). Nicotine regulation and cigarette smoking. *Clinical Pharmacology and Therapeutics*, **17**, 93–7.
17. Russell, M. A. H., Wilson, C., Feyerabend, C., and Cole, P. V. (1976). Effect of nicotine chewing gum on smoking behaviour and as an aid to cigarette withdrawal. *British Medical Journal*, **2**, 391–3.
18. Russell, M. A. H. (1979). Tobacco dependence: Is nicotine rewarding or aversive? In *Cigarette smoking as a dependence process*, National Institute on Drug Abuse Research Monograph, 23 (ed. N. A. Krasnegor), pp. 100–22. US Government Printing Office, Washington, DC.
19. Stolerman, I. P., Goldfarb, T., Fink, R., and Jarvik, M. E. (1973). Influencing cigarette smoking with nicotine antagonists. *Psychopharmacologia*, **28**, 247–59.
20. Schachter, S., Kozlowski, L. T., and Silverstein, B. (1977). Effects of urinary pH on cigarette smoking. *Journal of Experimental Psychology (General)*, **106**, 13–19.
21. Benowitz, N. L. and Jacob, P. (1985). Nicotine renal excretion rate influences

nicotine intake during cigarette smoking. *Journal of Pharmacology and Experimental Therapeutics*, **234**, 153–5.

22. Russell, M. A. H., Wilson, C., Patel, U. A., Feyerabend, C., and Cole, P. V. (1975). Plasma nicotine levels after smoking cigarettes with high, medium and low nicotine yields. *British Medical Journal*, **2**, 414–16.

23. Ashton, H., Stepney, R., and Thompson, J. W. (1979). Self-titration by cigarette smokers. *British Medical Journal*, **2**, 357–60.

24. Russell, M. A. H., Sutton, S. R., Iyer, R., Feyerabend, C., and Vesey, C. J. (1982). Long term switching to low-tar low-nicotine cigarettes. *British Journal of Addiction*, **77**, 145–58.

25. West, R. J., Russell, M. A. H., Jarvis, M. J., and Feyerabend, C. (1984). Does switching to an ultra-low nicotine cigarette induce nicotine withdrawal effects? *Psychopharmacology*, **34**, 120–3.

26. Sutton, S. R., Feyerabend, C., Cole, P. V., and Russell, M. A. H. (1978). Adjustment of smokers to dilution of tobacco smoke by ventilated cigarette holders. *Clinical Pharmacology and Therapeutics*, **24**, 395–405.

27. Benowitz, N. L. and Jacob, P. (1984). Daily intake of nicotine from cigarette smoking. *Clinical Pharmacology and Therapeutics*, **35**, 499–504.

28. Feyerabend, C., Ings, R. M. J., and Russell, M. A. H. (1985). Nicotine pharmacokinetics and its application to intake from smoking. *British Journal of Clinical Pharmacology*, **19**, 239–47.

29. Grunberg, N. E., Morse, D. E., Maycock, V. A., and Kozlowski, L. T. (1985). Changes in overwrap and butt length of American filter cigarettes: An influence on reported tar yields. *New York State Journal of Medicine*, **85**, 310–12.

30. Russell, M. A. H., Jarvis, M., Iyer, R., and Feyerabend, C. (1980). Relation of nicotine yield of cigarettes to blood nicotine concentrations of smokers. *British Medical Journal*, **280**, 972–6.

31. Ebert, R. V., McNabb, M. E., McCusker, K. T., and Snow, S. L. (1983). Amount of nicotine and carbon monoxide inhaled by smokers of low-tar low-nicotine cigarettes. *Journal of the American Medical Association*, **250**, 2840–2.

32. Benowitz, N. L., Hall, S. M., Herning, R. I., Jacob, P., Jones, R. T., and Osman, A. L. (1983). Smokers of low-yield cigarettes do not consume less nicotine. *New England Journal of Medicine*, **309**, 139–42.

33. Gori, G. B. and Lynch, C. J. (1985). Analytical cigarette yields as predictors of smoke bioavailability. *Regulatory Toxicology and Pharmacology*, **5**, 314–26.

34. Russell, M. A. H., Jarvis, M. J., Feyerabend, C., and Saloojee, Y. (1986). Reduction of tar, nicotine and carbon monoxide intake in low tar smokers. *Journal of Epidemiology and Community Health*, **40**, 80–5.

35. Doll, R. (1983). Prospects for prevention. *British Medical Journal*, **286**, 445–53.

36. Royal College of Physicians (1983). *Health or smoking?* Pitman, London.

37. Wald, N., Doll, R. and Copeland, G. K. E. (1981). Trends in tar, nicotine and carbon monoxide yields of UK cigarettes manufactured since 1934. *British Medical Journal*, **282**, 763–5.

38. Gori, G. B. and Bock, F. G. (eds) (1980). *Banbury Report 3: a safe cigarette?* Cold Spring Harbor, New York.

12

Isolating the role of nicotine in human smoking behaviour

M. J. JARVIS

Abstract

The evidence implicating nicotine as the major factor underlying tobacco use is reviewed. Nicotine acts as a prototypical drug of abuse in animal self-administration paradigms and its effect in enhancing performance in non-smokers points to its psychological rewarding properties. Administration of nicotine to smokers via gum or intravenously indicates a controlling influence on smoke inhalation and its replacement alleviates acute cigarette withdrawal symptoms, thus enhancing long-term cessation.

12.1. Introduction

The persistence of tobacco smoking within both societies and individuals is notorious. Historically, it has not been eliminated by the severest proscription, including torture and execution, and a substantial proportion of individuals who have suffered a heart attack,[1] or undergone laryngectomy,[2] or an operation for lung cancer[3] resume smoking again afterwards. Even now, 25 years after the first report of the Royal College of Physicians drew public attention to the scale of the health risks of smoking, less than half of ever-smokers of cigarettes in Britain give up permanently before the age of 65.[4]

It has been widely assumed that nicotine, as the major psychoactive substance in tobacco, underlies dependence on cigarettes. For example, it has been suggested that 'if it were not for the nicotine in tobacco smoke people would be little more inclined to smoke cigarettes than they are to blow bubbles or light sparklers'.[5] However, this view has not gone without challenge. Health educators have deliberately avoided mention of drug taking and have portrayed smoking as a social habit which millions have abandoned, usually with little difficulty. They have seen the concept of addiction as undermining this encouraging message, and as justifying rationalizations such as 'I can't stop: I'm an addict'. At the same time, nicotine researchers have drawn attention to serious limitations in the available evidence. In a 1977 paper reporting on an experiment which investigated the effects of intravenous nicotine on human smoking, the authors commented that 'tobacco smoking is generally regarded as a form of nicotine dependence, but the evidence for this is slender'. They also concluded that their own results did not support the nicotine dependence hypothesis.[6]

This paper will review briefly the evidence that implicates nicotine as the major factor underlying tobacco use, focusing particularly on recent findings in both man and animals which have greatly strengthened the case for regarding cigarette smoking as essentially nicotine self-administration. First, however, some comments on the areas of smoking behaviour that nicotine might account for may be in order.

12.2. What does the nicotine hypothesis postulate?

It is important to establish what the nicotine hypothesis of smoking implies. It does not, and could not, attempt to explain all the phenomena of smoking — for example, the gradient in prevalence and cessation by social class that has emerged over the past 20 years, or differences in the rate of uptake of the habit by men and women. It is no more than a truism to point out that, as with any other form of drug dependence, be it alcohol, opiates, or cocaine, social, psychological, political, and economic factors all have important parts to play, in addition to pharmacological influences. The claim is not that simple need for nicotine determines everything, but rather that nicotine is the primary reinforcer underlying tobacco use, and that if tobacco did not contain nicotine there would be no problem. Nicotine thus provides the background against which other factors play their part.

The nicotine model of smoking has sometimes been equated with an addiction (withdrawal relief) model, in which smoking is controlled by the negative consequences of not smoking. As the interval from the last cigarette increases, incipient withdrawal induces an aversive mood state which is terminated by smoking the next cigarette. An alternative view emphasizes the positive rewards of nicotine in terms of enhanced mood and performance. This has been labelled the 'psychological tool' model of smoking.[7] On this view, the motivation to continue smoking comes not from the need to stave off the consequences of withdrawal, but from a desire to experience positive effects which continue to be rewarding in their own right.

As Ashton and Stepney[7] have pointed out, these two views of nicotine's role, emphasizing its negative and positive effects respectively, are not mutually exclusive. Both types of motivation may be present, although to varying degrees, in all smokers, and both may lead to compulsive tobacco use. During cigarette deprivation, 'true' withdrawal effects on mood may be hard to distinguish from dysphoria caused by the absence of positively rewarding effects.

12.3.   Evidence for the role of nicotine

Patterns of tobacco use

The circumstantial evidence pointing to nicotine as critical to tobacco use is compelling. Tobacco has only ever been consumed in ways that permit absorption of pharmacologically significant amounts of nicotine into the blood. Absorption is through the lungs in the case of inhaled smoke, through the nasal mucosa with 'dry' snuff, and through the buccal mucosa with 'wet' snuff or non-inhaled smoking of pipes and cigars. Tobacco has never been eaten or made into a beverage, probably because extensive first pass metabolism in the liver would result in little nicotine reaching the systemic circulation unchanged. The only common factor between the various ways in which tobacco is taken is nicotine, since snuff and chewing tobacco are free of the combustion products generated when tobacco is smoked.

Comparison of regular daily 'dry' snuff takers and cigarette smokers has shown similar blood nicotine concentrations from habitual use, and a similar increase from a single pinch of snuff or single cigarette[8] (see Fig. 12.1). These findings strongly suggest a controlling influence of nicotine. It would be a remarkable coincidence if factors such as flavour, strength of tobacco, social influences, and so on just happened to produce similar blood nicotine concentrations from two such different behaviours. In a recent study of Swedish 'wet' snuff users we found substantial nicotine

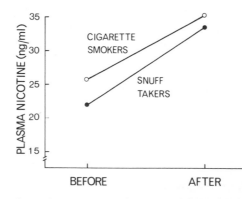

Fig. 12.1. Comparison of average trough and peak blood nicotine levels of regular snuffers (*n* = 7) and cigarette smokers (*n* = 13) before and after dosage during the afternoon of a day of usual usage. (From Russell *et al.*[8])

absorption from a single pinch placed between lip and gum. These subjects also reported feeling addicted to snuff and that they experienced frequent cravings for it (unpublished observations). Since both wet and dry snuff takers report strong dependence on their habit, it is clear that the presence of tar, carbon monoxide and other combustion products is not a prerequisite for the development of dependence on tobacco.

Nicotine as a primary reinforcer

A finding of wide generality is that drugs of abuse serve as reinforcers in drug self-administration paradigms. If nicotine is the main factor underlying compulsive use of tobacco, it would be expected that it would be a robust reinforcer of animal behaviour in controlled laboratory situations. For many years, however, it proved difficult to demonstrate consistent reinforcing effects of nicotine in animals under circumstances in which other drugs maintained high rates of responding.[9] Nicotine's apparent lack of reinforcing effects was puzzling, and represented a major weakness in the case for regarding tobacco smoking as essentially nicotine self-administration. The anomalous situation in which humans appear to manifest strong dependence on nicotine, but animals can hardly be induced to self-administer it at all, has now been resolved. Work by Goldberg, Spealman, and colleagues[10-12] has shown that the problem lay rather with the schedules of reinforcement applied than with the drug.

Self-administration behaviour was studied under complex intermittent schedules of reinforcement known as second order schedules. Under these, every ten lever pressing responses during a fixed interval of time produced a brief presentation of a light that had been associated with drug delivery; after the fixed interval had elapsed, the first ten responses completed produced both the light and an I.V. injection of nicotine (or, in other experiments, of cocaine). A timeout period followed during which the experimental chamber was in darkness.

Figure 12.2 shows representative cumulative response records from one squirrel monkey under these schedules. It can be seen that whereas saline produced no regular pattern of responding, intermediate dose levels of both nicotine and cocaine gave rise to high levels of responding. At the highest dose level (100 μg/kg/injection of nicotine or cocaine) patterns of responding were disrupted and rates of lever pressing decreased as the session progressed. The central nicotine antagonist mecamylamine blocked the reinforcing effects of nicotine and reduced responding to levels similar to those on saline.

Further experiments showed that, in addition to maintaining behaviour, under some conditions nicotine injection would lead to dose-related suppression of behaviour, and animals would work to postpone injections,

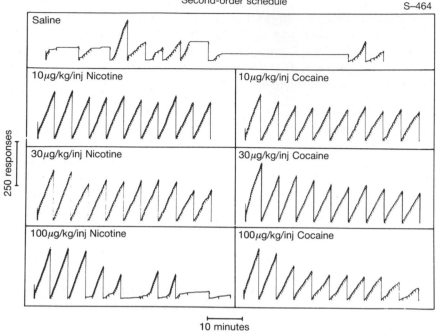

Fig. 12.2. Representative performances maintained by I.V. injections of nicotine or cocaine under the second-order fixed-interval schedule in a squirrel monkey (S-464). Abscissas: time; ordinates: cumulative responses. Diagonal marks show presentations of the 1-second visual stimulus. The recorder reset after each injection and did not operate during the timeout period. Each panel shows a complete record at the doses specified or when saline was substituted for the drugs. (From Spealman and Goldberg.[12] with permission.)

with highest rates of response to postponing the largest injections. Mecamylamine again blocked these effects.

Goldberg and Spealman's results, which were obtained in a variety of animal species, are of the greatest importance for establishing nicotine as a primary reinforcer. They show that under appropriate conditions nicotine is self-administered in a way closely similar to cocaine. The presence of both rewarding and punishing effects of nicotine indicates how important is the temporal patterning of nicotine's administration. The range of blood concentrations at which nicotine has rewarding effects is narrow and soon gives way at higher levels to aversive and toxic effects. With too high a dose or too frequent administration, rewarding effects are rapidly overtaken by aversive effects and behaviour suppressed. The key to demonstrating nicotine as a robust reinforcer seems to have had much to do with devising reinforcement schedules which ensured appropriately intermittent drug injections.

Direct evidence that nicotine acts similarly as a primary reinforcer in humans is not available. It would be ethically improper, as well as practically difficult, to study acquisition and maintenance of patterns of pure nicotine self-administration in subjects who had never smoked or used tobacco in other ways. However, if it could be shown that nicotine enhances mood and performance, this would provide indirect evidence for its reinforcing effects in humans. Studies of cigarette smokers given nicotine (or allowed to smoke) after overnight abstinence have suggested such performance enhancement,[13] but there are two main difficulties in interpreting these results. First, nicotine has generally been administered by smoking, so that nicotine effects and smoking effects are confounded; second, enhanced performance could represent relief of incipient withdrawal impairment rather than improvement in absolute levels of performance. Studies in non-smokers are straightforward to interpret, but have been hampered by the lack of acceptable means of delivering nicotine. Recently, more effective methods have become available, and it has been shown that a nasally administered pure nicotine solution substantially improves non-smokers' performance on a simple task of psychomotor speed, the enhancement being blocked by mecamylamine.[14] Effects of this kind could form the basis of the psychologically rewarding properties of the drug.

Effects of nicotine on human smoking behaviour

To demonstrate that nicotine exerts a controlling influence on human smoking behaviour, it is necessary to show that smokers respond to changes in nicotine availability with appropriate changes in their puffing and inhalation. This is the core issue underlying this symposium and is dealt with extensively elsewhere in this volume. The difficulty in interpreting the many studies involving brand switching or intake from self-selected brands of differing yields is that variation in nicotine yield is largely confounded with concurrent changes in tar and carbon monoxide yields. Thus, although there is no question that *smoke* intake is regulated, it is hard to exclude the possibility that factors other than nicotine are implicated.

This section will not attempt to cover any of the many studies which examine smoker compensation with varied yields. These studies can be suggestive, but no more. Rather, it will briefly detail the results of studies which have looked at the effects on smoking behaviour of giving nicotine by an alternative route, be it by nicotine chewing gum or intravenously.

Several double-blind cross-over studies have compared the effects of nicotine chewing gum with placebo on *ad libitum* smoking behaviour. When subjects were instructed to chew 10 pieces of gum per day, cigarette consumption was significantly reduced on the active gum by comparison with placebo, blood carboxyhaemoglobin levels were lower, and subjects also rated themselves as significantly more 'put off cigarettes'.[15] After taking gum and smoking as they felt inclined, plasma nicotine levels two minutes after a cigarette were not significantly different on the active and placebo gum (27.4 and 24.7 ng/ml, respectively). In addition, the concentration on the active gum did not differ from normal smoking levels (30.1 ng/ml). Since a blood concentration of about 10 ng/ml would be anticipated from 2 mg gum, this implies some inhibition of cigarette inhalation when chewing the gum.

Results of other studies have been similar, pointing to a decrease in the number of cigarettes smoked when chewing active gum,[16–18] and to a reduction in indices of puffing intensity.[17,18] Interestingly, in the study by Nemeth-Coslett and Henningfield, effects on self-rated desire to smoke were weaker than effects on smoking behaviour itself, pointing to some dissociation between objective and subjective indices of disposition to smoke.

The most direct way of testing nicotine's effects on smoking is to give nicotine intravenously and examine its effects on *ad libitum* smoking. This crucial experiment has been undertaken only three times. Lucchesi *et al.*[19] infused subjects with nicotine or saline during each of a number of experimental sessions. No effects were observed from a nicotine dose of 1 mg per hour, but there was a significant reduction in the number of cigarettes smoked in each of five subjects who received 22 mg of nicotine intravenously over a 6-hour period, and longer butts were also left in the nicotine condition as compared with saline. No measures of blood nicotine concentrations were available.

Kumar *et al.*[6] found no effect on measures of cigarette puffing of pulsed intravenous nicotine infusions designed to mimic intake from cigarettes. This contrasted with clear effects when comparable nicotine doses were given by inhalation of tobacco smoke. While the results of this study provided no evidence in support of the nicotine hypothesis, they are open to criticism, firstly on the grounds that measures of plasma nicotine or blood carboxyhaemoglobin (which would have indexed actual inhalation, rather than puffing) were not taken; and secondly, because there was an implicit demand for subjects to puff. The experiment was ostensibly designed to examine the effects of 'normal' smoking on a number of physiological measures, and a lit cigarette was continuously available in a cigarette holder a few inches from subjects' mouths. Thus, subjects may have puffed because this was what they thought they were supposed to do; whether they inhaled what they puffed is unclear.

The third study was completed recently in our laboratory and has not yet been published. In a double-blind cross-over design, after overnight abstinence from cigarettes, subjects were given, on one occasion, an intravenous pre-load of nicotine and, on another, saline. At the end of the infusion subjects were free to smoke as they wished. The results showed a powerful inhibitory effect of the nicotine preload on inhalation from the first post-infusion cigarette. As in the study by Nemeth-Coslett and Henningfield, self-reported desire to smoke was at variance with actual smoking behaviour, suggesting that nicotine's controlling influence was operating largely at a subconscious level.

Finally, in this section it is appropriate to note that intravenous nicotine self-administration has been studied in a small number of cigarette smokers.[20] Subjects consistently pressed a lever to receive injections of nicotine, but responded only erratically for saline. In the period immediately following nicotine self-administration sessions, there were significant reductions in numbers of cigarettes smoked and puffs taken.

Effect of nicotine on smoking withdrawal

As pointed out earlier, nicotine could influence tobacco use either through positive reward or through negative reinforcement. The latter refers to a hypothetical cycle in which abstinence from smoking, causing nicotine depletion, leads to adverse changes in mood, which are relieved by further nicotine intake. From this perspective, the argument is that people continue to smoke because the consequences of not smoking are unpleasant, and nicotine depletion is the cause of the unpleasantness. Evidence that nicotine replacement alleviates tobacco withdrawal symptoms would lend support to this view.

People experience a range of changes for the worse in mood and performance during cigarette withdrawal. Common psychological symptoms are irritability, restlessness, depression, hunger, difficulty concentrating, dizziness, and craving. These occur together with sufficient regularity to justify talking in terms of a withdrawal syndrome.[21] These symptoms also occur on cessation of smokeless tobacco use[22] or of nicotine chewing gum.[23] A number of short-term placebo-controlled trials have examined the effects of nicotine chewing gum on cigarette withdrawal.[24–26] The gum has consistently been found to alleviate total withdrawal discomfort, as well as a number of individual symptoms, particularly irritability and depression. Hughes and Hatsukami[27] reported that withdrawal relief was independent of whether their subjects could identify correctly whether they were assigned to active gum or to placebo, and hence could not reflect expectancy effects.

Some effects have generally not been relieved by the gum − including hunger and, more importantly, craving for cigarettes. This could imply either inadequate nicotine replacement or that some symptoms are due to loss of behavioural components of smoking. In a study in which smokers were randomly assigned to continue smoking their own brand or to smoke a very low yielding cigarette for a 10-day period, nicotine intakes on the ultra-low cigarette dropped by about 60 per cent, and there was a significant increase in hunger.[28] There were no other significant withdrawal effects and, in particular, there was little craving for subjects' usual brand. There was thus some evidence for nicotine depletion effects on the ultra-low yielding brand, but at the same time confirmation of a behavioural role in craving, as the lack of craving on the low yield cigarette contrasts with findings on similar levels of nicotine replacement from gum (although rate of absorption from the ultra-low cigarette would be faster than from the gum). The behaviour of smoking might reduce craving in two ways. It might do so because of previous association with the pharmacological action of nicotine (secondary reinforcement), or alternatively it may be that craving is directed specifically towards sensory or behavioural components of smoking. It has to be said that the lack of success of low-yield cigarettes in attracting market share suggests that purely behavioural factors in cigarette use make little contribution to long-term satisfaction.

Effect of nicotine replacement on smoking cessation

The effect of nicotine replacement on acute tobacco withdrawal symptoms suggests that this might also be a useful strategy in long-term smoking cessation. Nicotine chewing gum has been subjected to intensive study as an aid to cessation. Although results from minimal intervention trials have been inconsistent, with some positive outcomes,[29,30] being balanced by negative findings,[31,32] a number of clinic-based trials in which treatment with gum was combined with intensive psychological support have confirmed that the gum can have a substantial effect in enhancing cessation rates.[33-36] In two studies there was a doubling of validated 1-year cessation rates from about 15 per cent on placebo to 30 per cent on active gum,[33,36] and a recent meta-analysis which combined results from all available trials estimated a pooled 1-year sustained success rate on active gum of 23 per cent compared with 13 per cent on placebo.[37] In addition to the long-term outcome itself, process data have pointed to a specific effect of nicotine from the gum. The number of pieces of gum chewed was related to pre-treatment blood nicotine levels, and a small minority of patients developed dependence on active, but not placebo gum.[33]

The effects of nicotine gum on withdrawal and cessation are achieved through only modest nicotine delivery. Typically, blood levels of only one-third or less of those from normal smoking are maintained. That nicotine replacement to this level has specific effects strongly, if indirectly, argues for the key role of nicotine in the maintenance of cigarette smoking.

12.4. Summary and conclusions

For years it proved surprisingly difficult to demonstrate in a satisfactory way that nicotine underlies dependence on tobacco. Evidence that smokers compensate for changes in cigarette deliveries is suggestive, but by itself inconclusive, because nicotine delivery is confounded with that of other smoke components, and sensory factors may also be implicated. This chapter has pointed to the powerful evidence from the different ways in which tobacco is used and has summarized recent results from a number of areas which have gone a long way towards resolving the issue.

1. Nicotine acts as a prototypical drug of abuse in animal self-administration paradigms. Although ethical considerations preclude such studies in nicotine-naive humans, nicotine has been shown to have powerful effects in altering performance in non-smokers which could form a basis for the psychological rewarding properties of the drug.

2. Studies in which nicotine is administered to cigarette smokers by an alternative route (intravenously or by nicotine gum) indicate a controlling influence of nicotine on inhalation of cigarette smoke.

3. Nicotine replacement alleviates acute cigarette withdrawal symptoms and enhances long-term cessation.

In each of these instances nicotine exerts an effect which is independent of any other smoke component. Taken together with the extensive evidence of smoker compensation for changes in cigarette yields, these observations strongly implicate nicotine as the key factor underlying compulsive use of tobacco.

References

1. Burt, A., Illingworth, D., Shaw, R. T. D., Thornley, P., White, P. and Turner, R. (1974). Stopping smoking after myocardial infarction. *Lancet*, **i,** 304–6.
2. Himbury, S. and West, R. J. (1985). Smoking habits after laryngectomy. *British Medical Journal*, **291,** 514–5.
3. Davison, G. and Duffy, M. (1982). Smoking habits of long-term survivors of surgery for lung cancer. *Thorax*, **37,** 331–3.
4. Jarvis, M. J. and Jackson, P. H. (1988). Cigar and pipe smoking in Britain: implications for smoking prevalence and cessation. *British Journal of Addiction*, **83,** 323–30.
5. Russell, M. A. H. (1971). Cigarette smoking: natural history of a dependence disorder. *British Journal of Medical Psychology*, **44,** 1–16.
6. Kumar, R., Cooke, E. C., Lader, M. H., and Russell, M. A. H. (1977). Is

nicotine important in tobacco smoking? *Clinical Pharmacology and Therapeutics*, **21**, 520–9.

7. Ashton, H. and Stepney, R. (1982). *Smoking: psychology and pharmacology*. Tavistock Publications, London.

8. Russell, M. A. H., Jarvis, M. J., Devitt, G., and Feyerabend, C. (1981). Nicotine intake by snuff users. *British Medical Journal*, **283**, 814–7.

9. Henningfield, J. E. and Goldberg, S. R. (1983). Nicotine as a reinforcer in human subjects and laboratory animals. *Pharmacology Biochemistry and Behavior*, **19**, 989–92.

10. Goldberg, S. R., Spealman, R. D., and Goldberg, D. M. (1981). Persistent behaviour maintained by intravenous self-administration of nicotine. *Science*, **214**, 573–5.

11. Goldberg, S. R., Spealman, R. D., Risner, M. E., and Henningfield, J. E. (1983). Control of behaviour by intravenous nicotine injections in laboratory animals. *Pharmacology Biochemistry and Behavior*, **19**, 1011–20.

12. Spealman, R. D. and Goldberg, S. R. (1982). Maintenance of schedule-controlled behaviour by intravenous injections of nicotine in squirrel monkeys. *Clinical Pharmacology and Experimental Therapeutics*, **223**, 402–8.

13. Wesnes, K. and Warburton, D. M. (1983). Smoking, nicotine and human performance. *Pharmacology and Therapeutics*, **21**, 189–208.

14. West, R. J. and Jarvis, M. J. (1986). Effects of nicotine on finger tapping rate in non-smokers. *Pharmacology Biochemistry and Behavior*, **25**, 727–31.

15. Russell, M. A. H., Wilson, C., Feyerabend, C., and Cole, P. V. (1976). Effect of nicotine chewing gum on smoking behaviour and as an aid to cigarette withdrawal. *British Medical Journal*, **2**, 391–3.

16. Kozlowski, L. T., Jarvik, M. E., and Gritz, E. R. (1975). Nicotine regulation and cigarette smoking. *Clinical Pharmacology and Therapeutics*, **17**, 93–7.

17. Herning, R. I., Jones, R. T., and Fischman, P. (1985). The titration hypothesis revisited: nicotine gum reduces smoking intensity. In *Pharmacological adjuncts in smoking cessation*, NIDA Research Monograph 53 (ed. J. Grabowski and S. M. Hall) pp. 27–41. Department of Health and Human Services, Washington, DC.

18. Nemeth-Coslett, R. and Henningfield, J. E. (1986). Effects of nicotine chewing gum on cigarette smoking and subjective and physiologic effects. *Clinical Pharmacology and Therapeutics*, **39**, 625–30.

19. Lucchesi, B. R., Schuster, C. R., and Emley, G. S. (1967). The role of nicotine as a determinant of tobacco smoking frequency in man with observations of certain cardiovascular effects associated with the tobacco alkaloid. *Clinical Pharmacology and Therapeutics*, **8**, 789–96.

20. Henningfield, J. E., Miyasato, K., and Jasinski, D. R. (1983). Cigarette smokers self-administer intravenous nicotine. *Pharmacology Biochemistry and Behavior*, **19**, 887–90.

21. West, R. J. (1984). Psychology and pharmacology in cigarette withdrawal. *Journal of Psychosomatic Research*, **28**, 379–36.

22. Hatsukami, K. D., Gust, S. W., and Keenan, R. M. (1987). Physiologic and subjective changes from smokeless tobacco withdrawal. *Clinical Pharmacology and Therapeutics*, **41**, 103–7.

23. West, R. J. and Russell, M. A. H. (1985). Effects of withdrawal from long-term nicotine gum use. *Psychological Medicine*, **15**, 891–3.

24. West, R. J., Jarvis, M. J., Russell, M. A. H., Carruthers, M. E., and Feyerabend, C. (1984). Effect of nicotine replacement on the cigarette withdrawal syndrome. *British Journal of Addiction*, **79**, 215–9.
25. Schneider, N. G., Jarvik, M. E., and Forsythe, A. B. (1984). Nicotine versus placebo gum in the alleviation of withdrawal during smoking cessation. *Addictive Behaviors*, **9**, 149–56.
26. Hughes, J. R., Hatsukami, D. K., Pickens, R. W., Krahn, D., Malin, S., and Luknic, A. (1984). Effect of nicotine on the tobacco withdrawal syndrome. *Psychopharmacology*, **83**, 82–7.
27. Hughes, J. R. and Hatsukami, D. K. (1985). Short term effects of nicotine gum. In *Pharmacological adjuncts in smoking cessation*, NIDA Research Monograph 53 (éd. J. Grabowski and S. M. Hall) pp. 68–82. Department of Health and Human Services, Washington, DC.
28. West, R. J., Russell, M. A. H., Jarvis, M. J., and Feyerabend, C. (1984). Does switching to an ultra-low nicotine cigarette induce nicotine withdrawal effects? *Psychopharmacology*, **84**, 120–3.
29. Russell, M. A. H., Merriman, R., Stapleton, J., and Taylor, W. (1983). Effect of nicotine chewing gum as an adjunct to general practitioners' advice against smoking. *British Medical Journal*, **287**, 1782–5.
30. Hughes, J. R., *et al.* (1986). Efficacy of nicotine gum in general practice. Paper presented at Annual Meeting of the American Psychological Association, Washington DC, August 1986.
31. British Thoracic Society (1983). Comparison of four methods of smoking withdrawal in patients with smoking related diseases. *British Medical Journal*, **286**, 595–7.
32. Jamrozik, K., Fowler, G., Vessey, M., and Wald, N. (1984). Placebo controlled trial of nicotine chewing gum in general practice. *British Medical Journal*, **289**, 794–7.
33. Jarvis, M. J., Raw, M., Russell, M. A. H., and Feyerabend, C. (1982). Randomised controlled trial of nicotine chewing gum. *British Medical Journal*, **285**, 537–40.
34. Fagerstrom, K-O. (1982). A comparison of psychological and pharmacological treatment in smoking cessation. *Journal of Behavioral Medicine*, **5**, 343–51.
35. Hjalmarson, A. I. M. (1984). Effect of nicotine chewing gum in smoking cessation: a randomized placebo-controlled, double blind study. *Journal of the American Medical Association*, **252**, 2835–8.
36. Schneider, N. G., Jarvik, M. E., Forsythe, A. B., Read, L. L., Elliott, M. L., and Schweiger, A. (1984). Nicotine gum in smoking cessation: a placebo-controlled, double-blind trial. *Addictive Behaviors*, **8**, 253–61.
37. Lam, W., Sze, P., Sachs, H. S., and Chalmers, T. C. (1987). Meta-analysis of randomized controlled trials of nicotine chewing gum. *Lancet*, **ii**, 27–30.

13

The functional use of nicotine

D. M. WARBURTON

Abstract

The effects on improving mood and performance from the stimulating and tranquilizing effects of smoking, and specifically nicotine, are discussed and an argument presented for maintaining nicotine, but reducing tar yields of cigarettes. Comparison of a reduced-tar–maintained-nicotine cigarette with a conventional cigarette with similar nicotine, but higher tar yield, showed similar effects on mood and performance for a given nicotine yield.

13.1. Introduction

The typical view of international regulatory bodies has been the simple view that the total yields of cigarettes should be reduced.[1] However, one of the main conclusions of the *Third Report of the Independent Scientific Committee on Smoking and Health* of the United Kingdom[2] was that, although nicotine levels in general should continue to fall, there should also be available to the public some brands with low levels of tar and a proportionately higher nicotine yield, i.e. enhanced-nicotine cigarettes.

This Committee further recommended that: 'There should be careful monitoring of public acceptance of such brands and the extent of "compensation" . . .'. The acceptability of these reduced-tar cigarettes will depend in part on the extent to which they serve the needs of the smoker, i.e. they fit the Functional Model of Smoking.[2]

This chapter is concerned with the issue of whether any reduced-tar–maintained-nicotine cigarettes can satisfy the needs of the smoker in terms of the functional model of smoking.

13.2. The functional model of smoking

The Functional Model sees smoking as an important resource for the person. A resource refers not to what people do, but to what is available to them for managing their lives. The word 'managing' has been deliberately used instead of 'coping', since managing incorporates fewer normative connotations than coping. In other words, the Functional Model sees smoking as a resource for managing everyday life and not just for coping with problems.

As I have pointed out elsewhere,[3] there are many motives for smoking, but the Functional Model[4] does not require a single motive for all smokers. People may smoke for different reasons on different occasions. The level of smoking will reflect the personal function of smoking for the individual; the beneficial effects that smoking gives that person.

Smoking can be seen as being maintained by the personal control that smokers have over their psychological state, because of the very effective delivery of nicotine to the brain. Thus, smoking is a purposeful activity for smokers; it provides them with nicotine as a resource for managing their lives and virtually instant control of the person's psychological state.

Clues about the nature of the resources that are provided by the nicotine, have come from our work on comparative substance use. In this work subjects rated their experience of tranquillizers, amphetamine, and tobacco on a scale of zero to ten for their relaxing and stimulating properties. Tobacco has moderate stimulating and relaxing properties. It can be seen that subjectively it is much less stimulating than amphetamine (Fig. 13.1) and more stimulating than a tranquillizer while being more relaxing than amphetamine, but much less potent in this respect than a tranquillizer (Fig. 13.2).

We might expect that two important aspects of life, mood and work, might benefit from the stimulating and tranquillizing effects of smoking and each of these will be discussed in turn.

13.3. Smoking and mood

Watching a film of mutilation scenes is a standard way of producing a psychological and physiological stress response, inducing increases in heart rate and decreases in skin conductance. Smoking a cigarette decreases the

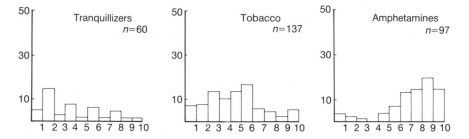

Fig. 13.1. Comparison of a prototypical tranquillizer and a prototypical stimulant with tobacco on the stimulation dimension.

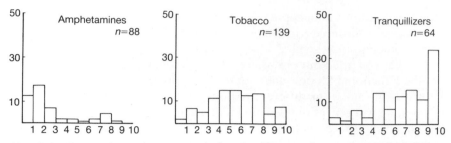

Fig. 13.2. Comparison of a prototypical tranquillizer and a prototypical stimulant with tobacco on relaxation.

subjective stress response and smoothes out the body's stress response.[5] Other recent studies in the laboratory have confirmed that cigarette smoking does give calming benefits to the smoker. It has been found that smoking reduced the annoyance[6] and physiological measures have shown that smoking dampens down the stress response to these noises.[6,7]

Smoking before testing enabled smokers to stand higher levels of electric shock than when they had not been allowed to smoke beforehand.[8] Smoking even helped smokers to be less angry when they knew that they were being cheated by another person in a gambling game.[9] Studies of the mood of smokers, during the smoking of a cigarette, show that the smoker feels calmer, more relaxed, more contented, friendlier, and happier, after a cigarette and are in a much better mood in general when smoking.[7]

In a study of the puff-by-puff mood effects of cigarettes, subjects completed a set of Bond–Lader mood scales[10] after each puff. This scale is a standard test that is used for assessing anxiolytics compounds. Subjects reported that they became calmer, more relaxed, more tranquil, more sociable, more friendly, more contented, and happier as they smoked the cigarettes (Fig 13.3 and 13.4). The changes on the 'relaxation' set of scales were curvilinear, while the puff-by-puff changes on the 'contentedness' were linear.

These laboratory studies of the calming effects of smoking are supported by the day-by-day experience of many smokers.[11–14] Surveys have found that 80 per cent of smokers say that they smoke more when they are worried, 75 per cent say that they light up a cigarette when they are angry, and 60 per cent feel that smoking cheers them up.[15,16]

These opinions suggest that smokers feel that smoking improves their mood and that this is an important reason for smoking.[17] If these beliefs of smokers were true, and the laboratory findings suggest that they are, then cigarette smoking is an important resource for smokers for calming them and reducing their feelings of anxiety and anger. Certainly, people increase their smoking when under stress.[8,17–19] It is not surprising that people miss

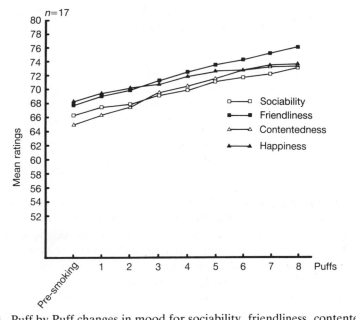

Fig. 13.3. Puff by Puff changes in mood for sociability, friendliness, contentedness and happiness.

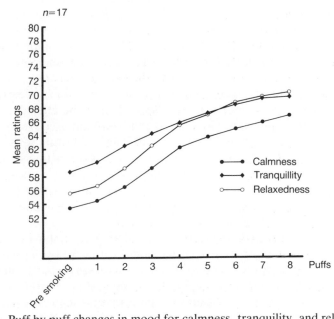

Fig. 13.4. Puff by puff changes in mood for calmness, tranquility, and relaxedness.

these benefits when they stop smoking. More anxiety and anger are two of the common complaints of people who cut down their smoking or give up.[20–23]

13.4.   Smoking, work, and productivity

It is common for smokers to say that smoking stimulates them, as smoking surveys have shown,[11,12,13,16] and, more specifically, that smoking helps thinking and concentration.[15,16]

Studies in the workplace have supported the idea of smoking being important for work. A survey of 2000 union representatives and managers in business, industry, and government revealed that smoking in the workplace can lift employee productivity and certainly does not retard it.[24] A similar survey among bank credit managers found smokers were 2.5 per cent more effective at their work.[25] A study of academic performance among university students revealed better examination marks and essay marks for smokers compared with non-smokers,[26] showing some association between smoking and academic success.

Laboratory studies have looked at a rather different type of performance, prolonged car driving. Using a driving simulator, researchers have found that fewer signals were missed by smokers smoking using a task which combined central guiding with peripheral signal detection.[27] Measures of steering efficiency, reaction time, brake light detection, and dial detection showed that smokers who were allowed to smoke during the 6-hour driving test were better on all aspects of performance.[28]

Other laboratory tests have looked at the effects of smoking on various components of these complex tasks. In a sustained reaction time task, reaction times normally slow down over the period of an hour, but smoking prevented the significant increase in reaction time which occurred when not smoking.[29,30] Researchers have also looked at the effects of smoking on concentration and have discovered that smoking a cigarette prevents the lapses in concentration which occur in long and boring tests[7,31,32] and more intense, rapid information processing.[33–36] It seems that smoking acts to reduce distraction by irrelevant signals.[31]

In a typical study from our laboratory,[35] we have studied information processing performance, both before and after smoking, and measured the speed and accuracy of processing. Smoking improved speed and accuracy above baseline levels, whereas either not smoking or smoking nicotine-free cigarettes resulted in a decline in speed and accuracy below baseline levels. Higher-nicotine cigarettes produced greater increases in speed and accuracy than lower delivery cigarettes (Fig. 13.5), suggesting that nicotine is the most important smoke constituent for performance improvement.

Studies using nicotine tablets have confirmed that nicotine is the most

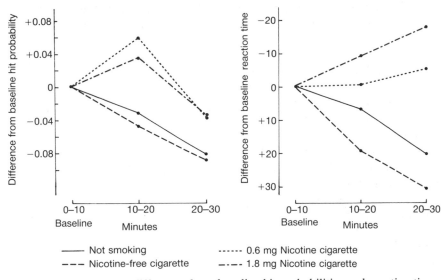

Fig. 13.5. Group mean difference from baseline hit probabilities and reaction times for each of the four conditions of experiment 2.

important ingredient of cigarette smoke for these improvements in performance. We gave a range of nicotine tablets to non-smokers before performing the rapid visual information processing task. Dr M. A. H. Russell showed that there was oral absorption of nicotine from the tablets. We found that the performance improvement was similar to that produced by smoking. From Fig. 13.6 it can be seen that nicotine improved the speed and accuracy of performance in the rapid information processing,[37,38] showing that nicotine was the active smoke constituent. Nicotine also increases tapping rate[39] and nicotine tablets prevent the lapses in attention which occur in the same long and boring tasks that are improved by smoking.[40] Nicotine reduces distraction by irrelevant stimuli just like smoking.[37]

We have also used the rapid visual information task to study the performance puff-by-puff during smoking a cigarette. In Fig. 13.7 and 13.8, it can be seen that, following the first puff, the probability of correct detections in the smoking conditions were significantly higher than in the non-smoking condition. A similar pattern occurs in the reaction-time data. Thus, a single bolus of nicotine significantly changes performance with the maximum performance being reached on the fourth puff.

Studies of the effects of smoking on learning and memory have shown that it improves this aspect of mental ability as well.[7,41,42] Studies of the effects of nicotine on learning and memory have shown that it improves recall.[41]

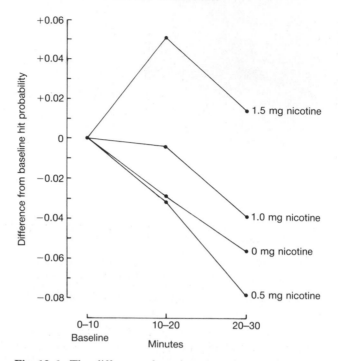

Fig. 13.6. The difference from baseline hit probabilities.

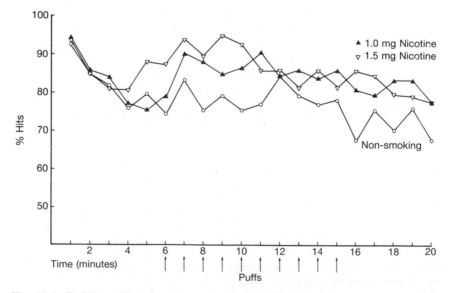

Fig. 13.7. Puff by puff changes in targets detected for varying yield cigarettes with low tar and middle tar.

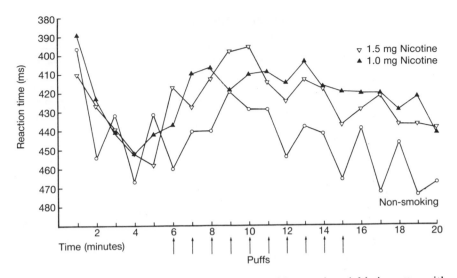

Fig. 13.8. Puff by puff changes in targets detected for varying yield cigarettes with low tar and middle tar reaction times.

A survey of smoking habits during the day[43] showed that peak smoking rates occurred during the working day. Given the above data, this suggests that, not only is smoking compatible with work, but it may be used as a resource to enhance work performance. It is not surprising that a United States Government Committee concluded that stopping airline pilots smoking 'may have a net adverse effect upon overall flight safety'.[44]

13.5. The derivation of nicotine from the cigarette

As part of these studies,[45] we have measured the nicotine retained in the cigarette filter and estimated mouth levels of nicotine (Table 13.1). Smokers obtained significantly more mouth nicotine from the three lower-yielding cigarettes than would be predicted from machine smoking, but less from the highest-yielding cigarette. In the same experiment, we also measured the levels of end-tidal carbon monoxide (ETCO) before and after smoking the cigarettes, in order to estimate the degree of smoke inhalation.

An inhalation intensity index, presented in Table 13.1 was obtained by dividing the increase in ETCO by the machine-smoked carbon monoxide yield. The inhalation intensity index of the lowest nicotine-yielding cigarette was significantly higher than that of the other three. The puffing intensity figures for the three lower-yielding cigarettes were not significantly different, indicating that the subjects had inhaled more intensely when smoking the lowest-nicotine cigarette and so obtained more nicotine.

Table 13.1. Smoke generation and smoke manipulation intensities during a performance study

Machine-smoked yield of nicotine (mg)	0.8	1.2	1.3	1.7
Generation intensity	115.1	116.6	112.3	*87.7*
Smoking intensity	*0.75*	0.51	0.41	0.49

In a performance study with varying nicotine-constant tar cigarettes, the ETCO values were correlated with the nicotine yields of the cigarettes showing the titration for nicotine by inhalation. Nicotine pre-load studies[45] have shown that it changes inhalation not puffing (Table 13.2).

During the performance task, subjects smoked cigarettes through a tube linked to a puff recorder. The cigarettes were varying nicotine-constant tar cigarettes (0.8, 1.2, and 1.35 mg of nicotine, and 11.5 mg of tar). Smoke generation measures on varying-nicotine–constant-tar cigarettes were similar, but subjects left longer intervals between puffs as nicotine yield increased (Fig. 13.9). Puff duplication on a smoking machine revealed that less tar was obtained from the higher nicotine cigarettes. These are important data for the reduced-tar–maintained-nicotine strategy because it indicates that subjects compensate for the lower nicotine yield by puffing harder and increase their level of mouth exposure to tar.

Mouth and plasma nicotine were studied while smoking a middle-tar and a low-tar cigarette, similar to those in our performance study. Mouth levels of nicotine were higher from the higher yield cigarette (Fig. 13.10). Blood samples were collected at 1-minute intervals during smoking. When measured puff-by-puff, increases in plasma nicotine were variable for most subjects. The only significant difference in puff-by-puff increases between the two cigarettes was during the first puff. Plasma nicotine increases were related to the yields of the two cigarettes (Fig. 13.11). Different asymptotes were reached towards the end of the cigarettes, but there was no significant difference, due to variation (Figs 13.12 and 13.13).

Table 13.2. Smoking manipulation and generation after nicotine pre-load

Tablet	Placebo	Placebo	1 mg Nicotine
Cigarette yield	1.4 mg	0.9 mg	0.9 mg
Puff number	16.2	16.5	16.8
Puff duration	1.8	1.8	1.8
Interpuff interval	25.5	22.8	22.4
Puffing intensity	133	117	122
Smoking intensity	0.43	0.61	0.43

Unpublished data from Wesnes, Pitkethley, and Warburton.

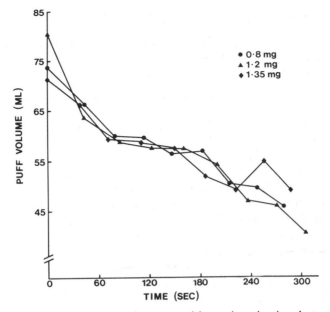

Fig. 13.9. Puff by puff profiles on cigarettes with varying nicotine, but constant tar.

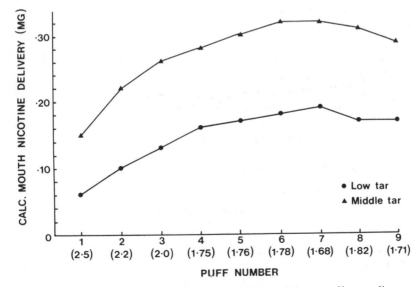

Fig. 13.10. Puff by puff mouth nicotine estimated from puff recordings.

D. M. Warburton

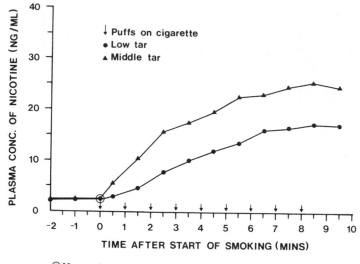

Fig. 13.11. Plasma levels of nicotine for a low tar and middle tar cigarette.

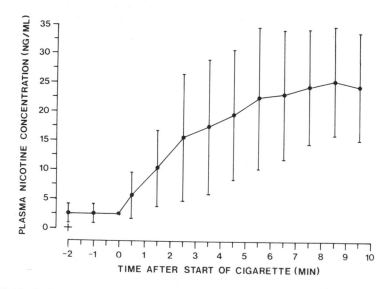

Fig. 13.12. Puff by puff nicotine levels for a middle tar cigarette showing the variability in plasma levels.

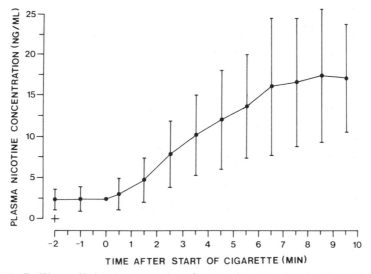

Fig. 13.13. Puff by puff nicotine levels for a low tar cigarette showing the variability in plasma levels.

The plasma nicotine changes can be related to performance and mood. Peak plasma levels of nicotine were not reached until the cigarette was finished. The maximum improvements during smoking were reached by the fifth puff and did not increase thereafter, while the mood measures increased throughout the cigarette. This finding suggests that the maximum effective dose for performance is less than that obtained from a whole cigarette and is always exceeded during smoking. In contrast, the plasma nicotine increase correlated more closely with mood; mood changes increased throughout the cigarette. This difference in pattern suggests that the maximum effective dose for mood change is higher than for performance. Thus, from the point of view of mood smokers are more likely to try and compensate.

13.6. Functional use of a reduced tar product

From the evidence that has been summarized above, smoking improves mood and performance and nicotine is the active ingredient in cigarette smoke that is responsible for these benefits. This evidence argues for the maintenance of nicotine levels in cigarettes. The smoke chemistry studies (chapter 7) suggest that any compensation that occurs does not result in increased tar exposure from the lower-tar–maintained-nicotine cigarette. The question then arises whether a reduced-tar cigarette can produce the same effects as conventional cigarettes with similar nicotine, but co-varying tar delivery.

We have compared a medium-nicotine–low-tar cigarette having a machine-estimated yield 1.4 mg of nicotine and 11.2 mg of tar (a tar to nicotine ratio of 8) with a medium-nicotine–medium-tar cigarette yielding 1.7 mg of nicotine and 16.9 mg of tar (a tar to nicotine ratio of 10). Our studies have been aimed at looking for similarities in the effects on mood and performance and the evidence shows that the two cigarettes produce similar effects.

In the mood study, we administered a set of Bond–Lader visual analogue scales, using a computer. Twenty female smokers assessed their own mood using the scale while smoking a cigarette. The subjects completed the set of scales after each puff on the cigarette. For both cigarettes, there were significant increases in calmness on the calm–excited scale over successive puffs (Fig. 13.14). Subjects also reported that they became more tranquil, more sociable, more friendly, more contented, more relaxed, and happier over successive puffs, and these changes were highly significant. However, no significant difference in the effects of the different cigarettes was found.

A comparison of the two cigarettes was also made using the rapid visual information processing task which is in routine use for testing the effects of smoking and nicotine on performance.[34–38] As we had found in many previous studies, smoking a cigarette prevented the decrease in detection of targets and increase in reaction time that occurs in the non-smoking condition. However, the enhanced nicotine cigarette did not differ from the conventional cigarette in its effects on the speed and accuracy of

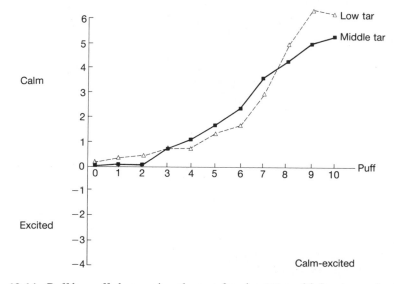

Fig. 13.14. Puff by puff changes in calmness for cigarettes with low tar and middle tar, but similar nicotine.

performance. There is marginally more improvement in accuracy with the reduced-tar cigarette and marginally more improvement in speed with the conventional cigarette, but these differences were not significant (Figs 13.15 and 13.16).

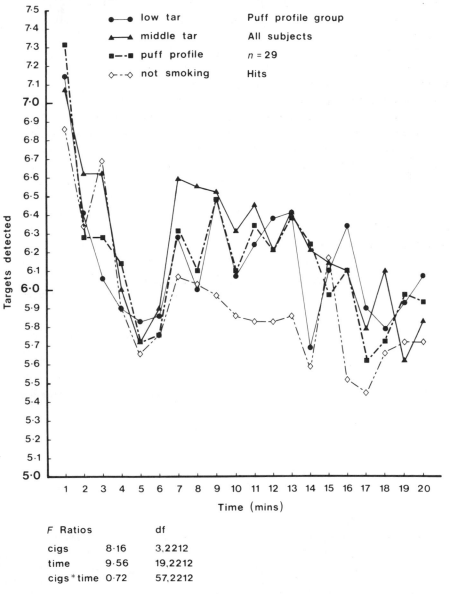

F Ratios		df
cigs	8·16	3,2212
time	9·56	19,2212
cigs*time	0·72	57,2212

Fig. 13.15. Puff by puff changes in targets detected for cigarettes with low tar and middle tar, but similar nicotine.

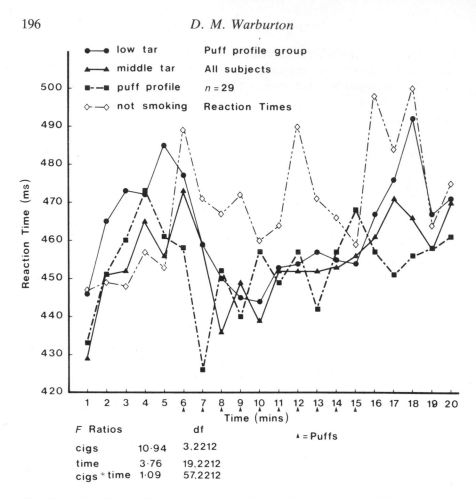

Fig. 13.16. Puff by puff changes in targets detected for cigarettes with low tar and middle tar reaction times.

13.7. Summary

In the introductory Section, it was stated that the acceptability of reduced-tar cigarettes will depend in part on the extent to which they fit the Functional Model of Smoking,[4] i.e. they serve the needs of the smoker. Two major psychological benefits that are sought by the smoker from smoking are improvements in mood and performance. It seems clear that both these benefits are due to nicotine and so a reduced-tar cigartte must maintain nicotine or else there will be compensation which will negate the tar reduction. A comparison of a reduced-tar–maintained-nicotine cigarette and a conventional cigarette with similar nicotine but higher tar shows that cigarettes with similar machine-estimated nicotine yields do have similar effects on these two aspects of psychological functioning. Thus, these cigarettes offer acceptable substitutes for conventional cigarettes.

References

1. WHO (1979). *Controlling the smoking epidemic.* World Health Organization, Geneva.
2. Independent Scientific Committee on Smoking and Health (1983). *Third Report.* HMSO, London.
3. Warburton, D. M. (1985). Nicotine and the smoker. *Reviews on Environmental Health*, **5**, 343–90.
4. Warburton, D. M. (1987). The functions of smoking. In *Tobacco smoke and nicotine: a neurobiological approach* (ed. W. R. Martin, G. R. Van Loon, E. T. Iwamoto, and D. L. Davies), pp. 51–62. Plenum Press, New York.
5. Gilbert, D. G. and Hagen, R. L. (1980). The effects of nicotine and extraversion on self-report, skin conductance, electromyographic, and heart responses to emotional stimuli. *Addictive Behaviors*, **5**, 247–57.
6. Woodson, P. P., Buzzi, R., Nil, R., and Battig, K. (1986). Effects of smoking on vegetative reactivity to noise in women. *Psychophysiology*, **23**, 272–82.
7. Mangan, G. and Golding, J. (1978). An 'enhancement' model of smoking maintenance? In *Smoking behaviour. Physiological and psychological influences* (ed. R. E. Thornton), pp. 115–26. Churchill Livingstone, Edinburgh.
8. Schachter, S. (1978). Pharmacological and psychologial determinants of smoking. In *Smoking behaviour. Physiological and psychological influences* (ed. R. E. Thornton), pp. 208–28. Churchill Livingstone, Edinburgh.
9. Cherek, D. R. (1981). Effects of smoking different dose of nicotine on human aggressive behaviour. *Psychopharmacology*, **73**, 339–45.
10. Bond, A. and Lader, M. (1974). The use of analogue scales in rating subjective feelings. *British Journal of Medical Psychology*, **47**, 211–18.
11. McKennell, A. C. (1970). Smoking motivation factors. *British Journal of Social and Clinical Psychology*, **9**, 8–22.
12. McKennell, A. C. (1973). *A comparison of two smoking typologies*, Tobacco Research Council, Research paper No. 12. Tobacco Research Council, London.
13. Tomkins, S. S. (1966). Psychological model for smoking behaviour. *American Journal of Public Health*, **56**, 17–20.
14. Tomkins, S. S. (1968). A modified model of smoking behaviour. In *Smoking, health and behaviour* (ed. E. Borgatta and R. Evans), pp. 165–8. Aldine, Chicago.
15. Russell, M. A. H., Peto, J., and Patel, U. A. (1974). The classification of smoking by factorial structure of motives. *Journal of the Royal Statistical Society*, **A137**, 313–33.
16. Warburton, D. M. and Wesnes, K. (1978). Individual differences in smoking and attentional performance. In *Smoking behaviour. Physiological and psychological influences* (ed. R. E. Thornton), pp. 19–43. Churchill Livingstone, Edinburgh.

17. Ashton, H. and Stepney, R. (1982). *Smoking: psychology and pharmacology.* Cambridge University Press, Cambridge.
18. Frith, C. D. (1971). Smoking behaviour and its relationship to the smoker's immediate experience. *British Journal of Social and Clinical Psychology*, **10**, 73–8.
19. Warburton, D. M. (1987). The functions of smoking. In *Tobacco smoke and nicotine: a neurobiological approach* (ed. W. R. Martin, G. R. Van Loon, E. T. Iwamoto, and D. L. Davies). Plenum Press, New York.
20. Guilford, J. S. (1966). *Factors related to successful abstinence from smoking.* American Institutes for Research, Pittsburgh.
21. Schechter, M. D. and Rand, M. J. (1974). Effect of acute deprivation of smoking on aggression and hostility. *Psychopharmacology*, **35**, 19–28.
22. Shiffman, S. M. (1979). The tobacco withdrawal syndrome. In *Cigarette smoking as a dependence process* (ed. N. A. Krasnegor), pp. 158–84. National Institute for Drug Abuse, Washington DC.
23. Shiffman, S. M. and Jarvik, M. E. (1976). Smoking withdrawal symptoms in two weeks of abstinence. *Psychopharmacology*, **50**, 35–9.
24. Response Analysis Corporation (1984). *Smoking and productivity in the workplace: overall report.* A Nationwide Survey Among: First Level Supervisors in Business and Industry; First Level Supervisors in Government; and Local Union Officials. Response Analysis Corporation. Princeton, New Jersey.
25. Dahl, T., Gunderson, B., and Kuehnast, K. (1984). *The influence of health improvement programs on white collar productivity.* University of Minnesota, Minneapolis.
26. Warburton, D. M., Wesnes, K., and Revell, A. (1984). Smoking and academic performance. *Current Psychological Research and Reviews*, **3**, 25–31.
27. Tarriere, H. C. and Hartemann, F. (1964). Investigations into the effects of tobacco smoke on a visual vigilance task. *Proceedings of the Second International Congress of Ergonomics*, 525–30.
28. Heimstra, N. W., Bancroft, N. R., and DeKock, A. R. (1967). Effects of smoking upon sustained performance in a simulated driving task. *Annals of the New York Academy of Science*, **142**, 295–307.
29. Frankenhaeuser, M., Myrsten, A.-L., Post, B., and Johansson, G. (1971). Behavioral and physiological effects of cigarette smoking in a monotonous situation. *Psychopharmacologia (Berlin)*, **22**, 1–7.
30. Myrsten, A-L. and Andersson, K. (1978). Effects of smoking on human performance. In *Smoking behaviour. Physiological and psychological influences* (ed. R. E. Thornton), pp. 156–67. Churchill Livingstone, Edinburgh.
31. Wesnes, K. and Warburton, D. M. (1978). The effects of cigarette smoking and nicotine tablets upon human attention. In *Smoking behaviour. Physiological and psychological influences* (ed. R. E. Thornton), pp. 115–26. Churchill Livingstone, Edinburgh.
32. Williams, G. D. (1980). Effect of cigarette smoking on immediate memory and performance in different kinds of smokers. *British Journal of Psychology*, **71**, 83–90.
33. Edwards, J. A., Wesnes, K., Warburton, D. M., and Gale, A. (1985). Evidence of more rapid stimulus evaluation following cigarette smoking. *Addictive Behaviours*, **10**, 113–26.

34. Wesnes, K. and Warburton, D. M. (1983). The effects of smoking on rapid information processing performance. *Neuropsychobiology*, **9**, 223–9.
35. Wesnes, K. and Warburton, D. M. (1984). The effects of cigarettes of varying yield on rapid information processing performance. *Psychopharmacology*, **82**, 338–42.
36. Wesnes, K. (1987). Nicotine increases mental efficiency: but how? In *Tobacco smoke and nicotine: a neurobiological approach* (ed. W. R. Martin, G. R. Van Loon, E. T. Iwamoto, and D. L. Davis), pp. 63–80. Plenum Press, New York.
37. Wesnes, K. and Revell, A. (1984). The separate and combined effects of scopolamine and nicotine on human information processing. *Psychopharmacology*, **84**, 5–11.
38. Wesnes, K. and Warburton, D. M. (1984). Effects of scopolamine and nicotine on human rapid information processing performance. *Psychopharmacology*, **82**, 147–50.
39. West, R. J. and Jarvis, M. J. (1986). Effects of nicotine on finger tapping rate in non-smokers. *Pharmacology, Biochemistry and Behaviour*, **25**, 727–31.
40. Wesnes, K., Warburton, D. M., and Matz, B. (1983). The effects of nicotine on stimulus sensitivity and response bias in a visual vigilance task. *Neuropsychobiology*, **9**, 41–4.
41. Warburton, D. M., Wesnes, K., Shergold, K., and James, M. (1985). Facilitation of learning and state dependency with nicotine. *Psychopharmacology*, **89**, 55–9.
42. Mangan, G. L. (1983). The effects of cigarette smoking on vigilance performance. *Journal of General Psychology*, **108**, 203–10.
43. Meade, T. W. and Wald, N. J. (1977). Cigarette smoking patterns during the working day. *British Journal of Preventative and Social Medicine*, **31**, 5–29.
44. National Institutes of Health Expert Panel (1978). *Cigarette smoking and airline pilots: effects of smoking and smoking withdrawal on flight performance.* A report on an expert panel of consultants. National Institutes of Health, Bethesda, Maryland.
45. Warburton, D. M., Wesnes, K., and Revell, A. D. (1984). Nicotine and the control of smoking behaviour. In *Smoking and the lung* (ed. G. Cumming and G. Bonsignore), pp. 217–32. Plenum, New York.

14

On the reduction of nicotine in cigarette smoke

DIETRICH HOFFMANN and ILSE HOFFMANN

Abstract

Reductions in the sales-weighted yields of tar and nicotine over the last 30 years have indicated a trend towards their differential reduction, and especially for the low and ultra-low tar cigarettes. Various means for the selective control of nicotine yields are discussed, including agricultural approaches, use of parts of the plant other than leaves (e.g. ribs or stems) as well as reconstituted tobacco sheet, varying tobacco blends, and manipulation by cigarette design.

14.1. Introduction

In this review on the reduction of nicotine and 'tar' we will refer exclusively to data obtained with machine-smoking of cigarettes under standardized laboratory conditions.[1-3] It is, of course, well known that these smoking techniques are not, or at best only to some degree representative of human smoking patterns. As is known for some time through the studies by Russell,[4] Herning et al.,[5] our own group,[6] and others, and as discussed at this Symposium by Rickert, Stephen, Adlkofer, and Benowitz,[7-10] the 'average smoker' tends to compensate for low smoke delivery of nicotine by smoking more intensely.

14.2. The commercial cigarette

In the USA, Canada and United Kingdom, the Federal Republic of Germany, Sweden, and several other countries, reduction of the sales-weighted tar and nicotine yields of commercial cigarettes over the last three decades has been substantial.[11-16] Figure 14.1 shows the decline of yields in the smoke of US cigarettes between 1955 and 1980, whereby sales-weighted averages of nicotine fell from 2.7 to 1.0 mg and tar from 38 to 14 mg, respectively, in each case a reduction of 63 per cent.

Although the trend towards a marked reduction of smoke yields from commercial cigarettes is universal, at least for most developed countries, the utilization of new tobacco types and of new methodologies in the

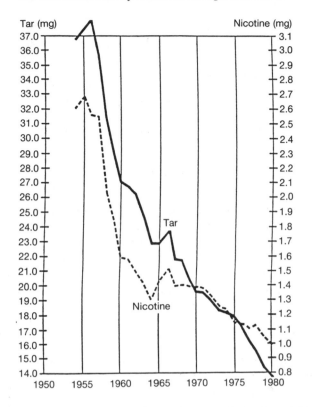

Fig. 14.1. US sales-weighted average tar and nicotine yields.

production of cigarettes can differ widely. Where tobacco blends are utilized for cigarettes, as for example in the US and in the Federal Republic of Germany, a major factor in reducing tar delivery has been in a proportional increase in the blend of Burley tobacco varieties. On the other hand, Burley varieties have found limited application in the UK, Finland, and Canada for a number of reasons, but primarily because smokers in these countries prefer cigarettes made of Bright tobaccos.

Figure 14.2 presents the data from Fig. 14.1 in graphic form and pinpoints the introduction of technical changes which have a profound influence on the sales-weighted average nicotine and 'tar' deliveries of commercial US cigarettes.[15]

14.3.   Changes in nicotine delivery

As shown in Fig. 14.2, between 1955 and 1980 the reduction in sales-weighted nicotine delivery of the US average cigarette parallels the tar

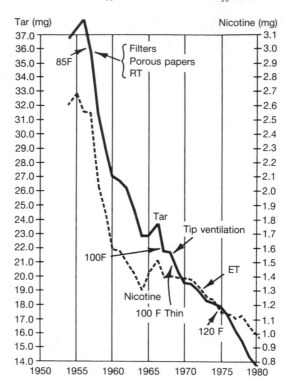

Fig. 14.2. US sales-weighted average tar and nicotine yields, (with indicators for changes in cigarette production).

reduction with a rather stable 1:14 ratio of nicotine to tar. Since 1981, the tar content has varied between 13.0 and 12.7 mg and the nicotine values have remained stable at 0.9 mg. However, according to the Federal Trade Commission report of January, 1985, the nicotine:tar ratio for the 16 high-tar cigarettes (>20 mg) is 1:15, for the 104 low-tar cigarettes (<11 mg) 1:12.6, and for the 40 ultra-low-tar cigarettes (<5 mg) 1:11.4 (Table 14.1[17]). These figures indicate the trend toward a differential in tar and nicotine reduction, especially for the low- and ultra-low-tar cigarettes.

According to Jarvis and Russell, the sales-weighted tar yields for cigarettes in the UK between 1972 and 1983 have decreased by about 29 per cent from 21 mg to about 15 mg, while the nicotine deliveries have remained stable varying only between about 1.2 and 1.35 mg per average cigarette (Fig. 14.3[14]). The nicotine to tar ratio for low-yield cigarettes (≤11 mg) was about 1:10 as compared to 1:11.7 for all other cigarettes (Fig. 14.4[14]).

The trends observed in the USA and especially in the UK, clearly

Table 14.1. Nicotine and tar in US cigarettes (1984)

Cigarettes		Nicotine (mg)	TPM (mg)	Nicotine:TPM ratio
No.	Types			
16	High tar	1.2–2.1 (1.54)	20–28 (23.4)	1:14
104	Low tar	0.1–0.9 (0.56)	1–11 (7.0)	1:12.6
40	Ultra-low tar	0.1–0.5 (0.33)	1–5 (3.7)	1:11.4

FTC (1985).[17]

demonstrate the intention of the manufacturers to reduce the tar yields of cigarettes to a greater extent than the nicotine yields. This time trend in the smoke yields of cigarettes has been called by Russell 'the low-tar, medium-nicotine approach'.[4]

14.4.   Means for selective reduction of the smoke yields of nicotine

The possibilities for the reduction of nicotine in cigarette smoke have been reviewed in 1975 by Kuhn and Klus.[18] Figure 14.5 summarizes these possibilities beginning with changes in agricultural approaches, e.g. selecting and breeding varieties that yield less nicotine in the smoke as well as changes in leaf-curing and fermentation. Modifications in the make-up of the blend by using expanded laminae and reconstituted tobacco sheets were also considered more than 10–15 years ago. In addition, Kuhn and Klus

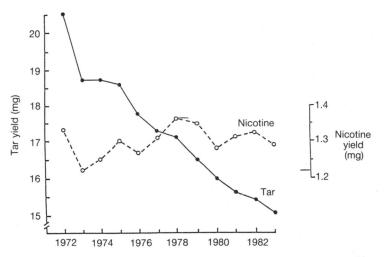

Fig. 14.3. Tar and nicotine yields of UK cigarettes, 1972–983.[14]

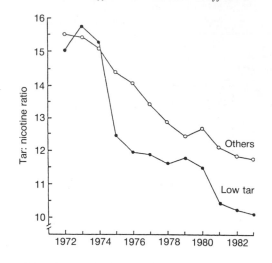

Fig. 14.4. Sales-weighted tar:nicotine ratios in low-tar (<11 mg) and other
cigarettes, 1972–1983.[14]

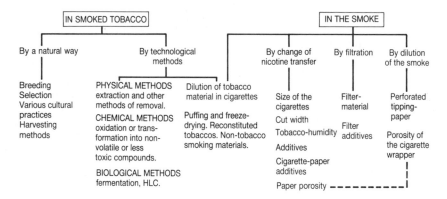

Fig. 14.5. Possibilities for reduction of smoke nicotine.[18]

cited several methods for reducing the nicotine yields in cigarette smoke
such as using more porous cigarette paper, selective filtration and smoke
dilution, primarily utilizing perforated filter tips.

   The primary methods for selectively reducing nicotine in the smoke are
those based on selection of tobacco leaves with low nicotine content. Table
14.2 presents recent data for nicotine concentrations in various tobacco
types,[19] demonstrating the possibility of controlling the nicotine yield in
the smoke by selecting specific Bright or Burley varieties that are low or
high in the major *Nicotiana* alkaloids.

Table 14.2. Alkaloid content of various tobacco brands (mg/kg, dry basis)

Alkaloid	Dark commercial tobacco		Burley	Bright	Kentucky ref. (IR1)[6]
	A	B			
Nicotine	11 500	10 000	15 400	12 900	21 100
Nornicotine	550	200	630	210	630
Anatabine	360	380	570	600	930
Anabasine	140	150	90	150	190
Cotinine	195	140	90	40	80
Myosmine	45	50	60	30	85
2,3'-Dipyridyl	100	110	30	10	30
N'-Formyl-nornicotine	175	210	140	40	100

Piade and Hoffmann (1980).[19]

Another approach for modulating nicotine yields in the smoke is that of utilizing laminae from the leaves harvested from various stalk positions. Table 14.3 presents data from a study completed in co-operation with the US Department of Agriculture. The leaves from four Bright tobacco varieties were grouped according to eight stalk positions. Position 1 was the lowest level of leaves on the stalk and position 8 the highest. The leaves from each of the eight stalk positions were independently cured, processed, and made into 85-mm non-filter cigarettes. The nicotine and tar yields in the smoke of cigarettes prepared in this manner were determined for each of the four Bright tobacco varieties as listed in Table 14.3. Clearly, nicotine and tar yields in the cigarette smoke increase with ascending stalk position and the nicotine:tar ratios decrease.[20] LN-38 is a low-yield Bright tobacco variety which did not show a significant change in the nicotine:tar ratio, but demonstrated the possibility of reducing nicotine yields by selecting special tobacco varieties. We also studied the smoke yields of nicotine and tar from cigarettes made with leaves of four Burley tobacco varieties. Again, with ascending stalk position of the leaves, nicotine and tar in the smoke increased and the nicotine:tar ratio decreased.[21] Although this approach to a control of the nicotine yields in smoke is intriguing, its practical application may have certain limits.

The most widely used technique to control tar and nicotine delivery in cigarette smoke is the incorporation into the tobacco blend of ribs and stems either in open form or as reconstituted tobacco sheets.[22] Ribs and stems are relatively lower in nicotine and alkaloid content than the laminae of the same leaves.[23]

Possibilities for a selective control of nicotine yields in the smoke include

Table 14.3. Yields of nicotine and TPM of cigarettes made of leaves from different stalk positions of Bright tobacco varieties*

Bright variety	Stalk position	Nicotine (mg)	TPM (mg)	Nicotine:TPM ratio
SC–58	L.L.	0.79	18.2	1:23
	H.L.	6.70	44.8	1: 6.7
Coker–139	L.L.	0.42	17.3	1:41
	H.L.	2.98	43.4	1:14.5
NC–95	L.L.	0.77	17.2	1:22.3
	H.L.	3.55	35.2	1: 9.9
LN–38	L.L.	0.19	17.06	1:90
	H.L.	0.46	33.15	1:72

*Four cigarettes, each containing 1000 mg tobacco; TPM = Total tar minus nicotine minus water; L.L., low stalk position; H.L., high stalk position.
Rathkamp *et al.* (1973).[20]

varying cigarette papers or filter tips. Owens reported data showing significant and differential variation in smoke yields when the porosity of the cigarette paper is changed (Fig. 14.6[24]). Conventional filter tips do not usually enable selective control of nicotine in the smokestream. This applies to cigarettes made with Bright tobaccos as well as those made with tobacco blends (Fig. 14.7[25]). The smoke of these cigarettes is weakly acidic;[26] thus, nicotine is protonized and occurs only in the particulate matter of the smoke. This renders it unamenable to selective filtration.

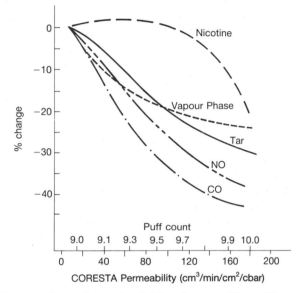

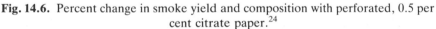

Fig. 14.6. Percent change in smoke yield and composition with perforated, 0.5 per cent citrate paper.[24]

In the case of cigarettes which produce slightly basic smoke such as the French black cigarettes, a significant proportion of the nicotine is present in free form and, therefore, appears in the vapor phase from where it can, to some extent, be selectively removed by appropriate filter materials in the tip.[26]

Cigarettes with highly perforated filter tips allow reduction of tar in the smokestream to a further extent than nicotine as was already shown in 1974 by Norman (Fig. 14.8[27]). Conversely, highly perforated filter tips reduce nicotine to a lesser extent than tar.

14.5. Discussion

This paper deals primarily with the observation that the mainstream smoke yields of low-yield cigarettes now available in many developed countries

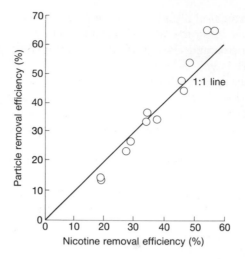

Fig. 14.7. Comparison of removal efficiencies.[25]

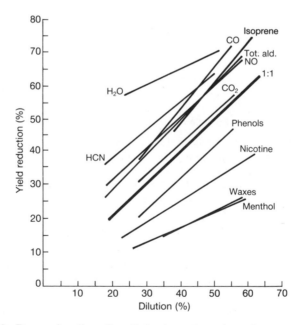

Fig. 14.8. Regression lines for all the investigated smoke components.[27]

reflect a greater reduction of tar than of nicotine. It is likely that this is by design.

The discussion focuses on possibilities for reducing nicotine yields in the smoke of cigarettes even to the point where the nicotine:tar ratio is changed in favour of nicotine. However, the 'low-tar, medium-nicotine approach' should not be endorsed by the academic community since nicotine is not only the major habituating factor in tobacco products, but also gives rise to the most abundant, highly potent carcinogenic tobacco specific N-nitrosamines.[28,29]

14.6. Summary

The last few decades have witnessed in many developed countries a gradual and highly significant reduction in the smoke yields of tar and nicotine of commercial cigarettes. However, during the last 5–10 years the yields of nicotine in the smoke of cigarettes in the UK and the USA have remained rather stable whereas tar in the smoke of commercial cigarettes has been further reduced. The paper reviews also the various methodologies for the reduction in cigarette smoke of nicotine, the major habituating agent in tobacco products.

Acknowledgements

We thank Mrs Bertha Stadler for assistance in the preparation of this manuscript. Our studies on tobacco are supported by Grant No. CA–29580 from the US National Cancer Institute.

References

1. Pillsbury, H. G., Bright, C. C., O'Connor, K. J., and Irish, F. W. (1969). Tar and nicotine in cigarette smoke. *Journal—Association of Official Analytical Chemists*, **52**, 458–62.
2. Rothwell, K. and Grant, C. A. (1974). *Standard methods for the analysis of tobacco smoke* (2nd edn). Tobacco Research Council, London.
3. Brunnemann, K. D., Hoffmann, D., Wynder, E. L., and Gori, G. B. (1976). Determination of tar, nicotine, and carbon monoxide in cigarette smoke. A comparison of international smoking conditions. DHEW Publication number (NIH) 76–1221, pp. 441–9.
4. Russell, M. A. H. (1977). Smoking problems: an overview. In *Research on smoking behavior*, National Institute on Drug Abuse Research Monograph Series, Vol. 17, (ed. Jarvik, M. E. *et al.*), pp. 13–34. DHEW Publication No. [ADM] 78–581. Washington DC.
5. Herning, R. I., Jones, R. T., Bachman, J., and Mines, A. H. (1981). Puff volume increases when low-nicotine cigarettes are smoked. *British Medical Journal*, **283**, 187–9.

6. Haley, N. J., Sepkovic, D. W., Hoffmann, D., and Wynder, E. L. (1985). Cigarette smoking as a risk for cardiovascular disease, Part VI, Compensation with nicotine availability as a single variable. *Clinical Pharmacology and Therapeutics*, **38**, 164–70.

7. Rickert, W. (1986). Do machine yields reflect yields to smokers? ISCSH Symposium, November 18–20. (This volume, chapter 7.)

8. Stephen, A. (1986). Summary of results of compensation studies. ISCSH Symposium, November 18–20. (This volume, chapter 8.)

9. Adlkofer, F. (1986). Consistency of nicotine intake in smokers of cigarettes with varying yields of tar and nicotine. ISCSH Symposium, November 18–20. (This volume, chapter 9.)

10. Benowitz, N. (1986). Dosimetric studies of compensatory cigarette smoking. ISCSH Symposium, November 18–20. (This volume, chapter 10.)

11. Weber, K. H. (1976). Recent changes in tobacco products and their acceptance by the consumer. *Proceedings of the 6th International Tobacco Science Congress*, Tokyo, pp. 47–63.

12. Government Chemist (1981). *Report of the Government Chemist 1980*. HMSO, London.

13. Wald, N., Doll, R., and Copeland, G. (1981). Trends in tar, nicotine, and carbon monoxide yields of U.K. cigarettes manufactured since 1934. *British Medical Journal*, **282**, 763–5.

14. Jarvis, M. J. and Russell, M. A. H. (1985). Tar and nicotine yields of U.K. cigarettes, 1972–1983: sales-weighted estimates from non-industry sources. *British Journal of Addiction*, **80**, 429–34.

15. Norman, V. (1982). Changes in smoke chemistry of modern day cigarettes. *Recent Advances in Tobacco Science*, **8**, 141–77.

16. International Agency for Research on Cancer (1986). *Tobacco smoking*, IARC Monograph on the evaluation of carcinogenic risk of chemicals to man, Vol. 38. IARC, Lyon, France.

17. US Federal Trade Commission (1985). Tar, nicotine and carbon monoxide of the smoke of 207 varieties of domestic cigarettes. US Federal Trade Commission, Washington, DC, January, 1985.

18. Kuhn, H. and Klus, H. (1976). *Possibilities for the reduction of nicotine in cigarette smoke*. Proceedings of the 3rd World Conference on Smoking and Health. Volume 1. Modifying the risk for smokers, (ed. Wynder, E. L., Hoffmann, D. and Gori, G. B.). DHEW Publication number (NIH) 76–1221, pp. 463–94.

19. Piade, J. J. and Hoffmann, D. (1980). Chemical studies on tobacco smoke. LXVII. Quantitative determination of alkaloids in tobacco by liquid chromatography. *Journal of Liquid Chromatography*, **3**, 1505–15.

20. Rathkamp, G., Tso, T. C. and Hoffmann, D. (1973). Smoke analysis of cigarettes made from Bright tobaccos differing in variety and stalk positions. *Beitrage zur Tabakforschung*, **7**, 179–94.

21. Tso, T. C., Chaplin, J. F., Adams, J. D. and Hoffmann, D. (1982). Simple correlation and multiple regression among leaf and smoke characteristics of Burley tobaccos. *Beiträge zur Tabakforschung*, **11**, 141–9.

22. Halker, M. M. and Ito, T. I. (1978). Effect of tobacco reconstitution and expansion processes on smoke composition. *Recent Advances in Tobacco Science*, **4**, 113–32.

23. Neurath, G. and Ehmke, H. (1964). Untersuchungen über den Nitratgehalt des Tabaks. *Beitrage zur Tabakforschung*, **2**, 333–44.
24. Owens, W. F., Jr (1978). Effect of cigarette paper on smoke yield and composition. *Recent Advances in Tobacco Science*, **4**, 3–24.
25. Keith, C. H. (1978). Physical mechanisms of smoke filtration. *Recent Advances in Tobacco Science*, **4**, 25–45.
26. Brunnemann, K. D. and Hoffmann, D. (1974). The pH of tobacco smoke. *Food and Cosmetics Toxicology*, **12**, 115–24.
27. Norman, V. (1974). The effect of perforated tipping paper on the yields of various smoke components. *Beiträge zur Tabakforschung*, **7**, 282–7.
28. Hoffmann, D. and Hecht, S. S. (1985). Perspectives in Cancer Research. Nicotine-derived N'-nitrosamines and tobacco-related cancer: current status and future directions. *Cancer Research*, **45**, 934–44.
29. Hoffmann, D. (1986). Nicotine, a tobacco-specific precursor for carcinogens. Presented at ISCSH Symposium, November 18–20. (This volume, chapter 3.)

15

The possible role of factors other than nicotine in compensatory smoking

FRANK A. FAIRWEATHER

Abstract

Numerous factors interrelate in the smoker's appreciation of a cigarette and whilst nicotine has always been considered to be the main stimulus, it is clear it is not the only factor determining smoking behaviour, including compensatory smoking. Evidence for the role played by factors such as taste, flavour, and the physical design of a cigarette remains poor, but equally no study has shown unequivocally that nicotine alone determines smoking behaviour.

15.1. Introduction

The role played by nicotine in compensatory smoking has been the subject of extensive recent research with the consensus view that smokers on average titrate nicotine to some extent (to maintain a personal optimal level available to the operative brain centres), but that other factors are also important determinants. The attention paid to these other factors has been somewhat piecemeal with a distinctly limited literature, but they must be taken into account since the complexity of smoking is the result of the smoker adjusting his behaviour to satisfy his requirements, but whose behaviour is affected directly by the product. This paper will discuss the limited information currently available on the possible role of factors other than nicotine in compensatory smoking.

15.2. The act of smoking

The very act of smoking a cigarette is a complex procedure which involves the drawing in of a complex smoke mixture into the mouth. This so-called 'main-stream' of smoke is accompanied by some leakage of 'wasted smoke' via the nostrils. There follows manipulation of this smoke within the mouth so that a bolus of smoke, adequately diluted by air may be presented, at least in part, to the upper respiratory tract and lungs for inhalation and absorption. This is illustrated in Fig. 15.1.

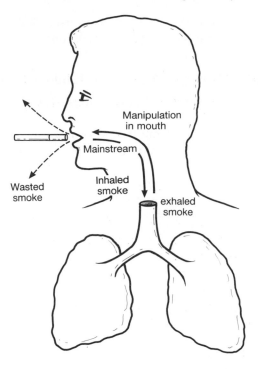

Fig. 15.1. Phases of smoke acquisition. (a) Puffing behaviour. (b) Mouth manipulation. (c) Inhalation.

The difference between mouth intake of smoke and subsequent inhalation cannot be overstressed since they are two independent steps, and for the smoker the most important final outcome is believed to be the dose of smoke constituents actually absorbed although it cannot be disputed that other factors play a role in the enjoyment of the act.

15.3. Compensation – a broad definition

The term 'compensation' has not been adequately defined in the literature. Hence, workers adopt their own definitions, not only as far as specific formulae are concerned, but also in terms of the smoking behaviour measurements that are made and the experimental designs adopted.

Essentially, 'compensation' is concerned with measuring changes in *smoking behaviour* and the *resultant intake of smoke substances* that may occur when a smoker changes from one product to another. Compensation is a relative term which does not necessarily indicate the smoker at the lower level of exposure is attempting to achieve the higher exposure level.

No underlying mechanism should be inferred from the term; it could as easily be called 'de-compensation' as 'compensation'.

15.4.   Factors that influence cigarette smoke constituents in the body

Tables 15.1–15.3 list the variables, in both the cigarette and the smoker, which will affect the availability or intake of cigarette smoke constituents in the body and hence may play some role in compensation. Some of these factors may account for why compensation occurs whereas others may describe, in part, how it occurs.

Cigarette construction factors are discussed fully in the Banbury Report[1-6] and some of the behavioural factors by Rawbone.[7] Indeed, it is timely to note here that the evidence for the measurement and quantification of dose of smoke in the smoker is vast, and some approaches are considered more relevant than others, especially bearing in mind the factors seen in Fig. 15.1. Rawbone[7] outlines some of these available techniques and their applications.

Although here we are not considering nicotine, it must be said that there is a widely held belief that cigarette smoking is a form of nicotine dependence and that smokers manipulate their smoking behaviour to obtain an optimum level of nicotine in the blood. Over 90 per cent of all cigarette smokers inhale the acid smoke into the lungs, thus allowing absorption of the alkaloid into the blood stream and its extremely rapid transfer to the brain. Inhalation is not a natural behaviour and has to be learnt at some considerable discomfort. It is not necessary for the olfactory, oral, or manipulative satisfaction of smoking. However, the level of nicotine in the blood attained by the smoker is frequently large enough to have considerable pharmacological effects.

Table 15.1. Factors that influence smoke constituents in the body: cigarette construction

Tobacco composition
   Genetic variety
   Agriculture: curing, fertilization, pesticides
   Parts of plant used
   Flavouring and additives
   Moisture content

Physical design
   Filtration: type of filter, ventilation
   Tobacco: density, expansion, consistency
   Cigarette: length, diameter
   Paper: porosity, additives

A growing number of studies, however, are providing convincing evidence that nicotine titration is not the over-riding, still less the exclusive determinant of smoking behaviour. Nicotine undoubtedly has a significant role in the tobacco habit, but it may be smaller than often supposed with such factors as taste, perception, smell, sociability, and manual manipulation, recently somewhat submerged by the pharmacological lobby (though not by the industry), being of at least equal importance.

Table 15.2 Factors that influence cigarette smoke constituents in the body: individual behavioural parameters

Brand choice

Brand history

Amount smoked
  Number of cigarettes per day
  Number of puffs per cigarette
  Puff distribution down cigarette

Puffing behaviour
  Puff volume
  Puff duration
  Interpuff interval
  Puff shape

Inhalation behaviour
  Inhalation timing in relation to puff
  Inhalation volume
  Duration of inhalation
  Breath-holding
  Normal respiratory parameters
  Waste smoke

Manner of holding cigarette
Passive exposure
Social/situational factors

Table 15.3. Factors that influence cigarette smoke constituents in the body: individual physiological factors

Smoking history
Environmental interactions
Physical status: height, weight, fitness
Health status: metabolic disorders, respiratory problems
Metabolic rates: normal or drug induced changes
Physical defences against toxins
Relative and variable flow rates of biofluids
Interactions between smoke constituents and biochemicals

15.5. Cigarette design

Cigarette design factors that may affect compensatory smoking can be
conveniently separated into two categories—the physical characteristics of
the product (especially pressure drop and filter design) and the chemistry
of the smoke (not only nicotine, tar, and carbon monoxide yield, but also
the 'sensory-active' substances; the latter may or may not correlate with
nicotine, tar, or carbon monoxide).

In terms of physical characteristics the cigarette pressure drop has been
shown to influence specifically puff volume. Rawbone[8] reports that as
pressure drop increases the volume of the individual puffs decreases. He
concludes that smokers may be standardising on the amount of physical
effort required to take a puff. Clearly, nicotine is not a prime factor,
pressure drop itself being controlled by such factors as filter design,
cigarette paper porosity, and filler (tobacco) characteristics. Dunn[9] found
puff volume to increase when draw resistance was decreased, but more
importantly, though mouth level of nicotine was greatly increased alveolar
carbon monoxide levels decreased, indicating the importance of the estimate of inhalation.

In terms of smoke chemistry, it is clear that cigarettes differ from brand
to brand not only in nicotine, tar, and carbon monoxide, but also in taste
and flavour (sensory active substances). Indeed, of all the possible factors
other than nicotine which may contribute to compensatory smoking, most
attention has probably been paid to these sensory substances, despite
problems in their definition. Cain[10] has discussed the possible importance
of stimulation of the common chemical sense by cigarette smoke. This
latter sensory factor mediates pungency, warmth, cold, pain, piquancy,
acidity, and astringency. Cain points out that sensory factors may serve as
an immediate sign to the smoker of effects which follow shortly after the
puff, e.g. relaxation, reduced anxiety, and so provide an important cue to
the positive features of the product. He believes that the smoker, in terms
of less hazardous smoking, could be 'nudged very gradually in the direction
of a good tasting cigarette'. Clearly, tobacco research into the area of taste,
flavour, and common chemical sense stimulation is in its infancy. Indeed,
the relative contribution made by nicotine itself by virtue of any possible
intrinsic sensory activity, when compared with other substances, is not yet
understood.

In a recent study the importance of upper respiratory tract sensation has
been examined by Rose and colleagues[11] who employed local anaesthesia
(lidocaine) in an attempt to block sensory responses to cigarette smoke.
They were able to conclude that 'sensory cues accompanying inhalation of
cigarette smoke are important determinants of immediate smoking satisfaction'. Rose *et al.* consider that the nicotine boli view[12] of the 'uniquely

desirable qualities of cigarettes to a smoker' may not be the total story and that an 'alternative account for the enjoyment and satisfaction' is the 'conditional reinforcing effects of the sensory feedback accompanying inhalation'. Once again, however, the results do not allow one to conclude that nicotine is totally inactive at the sensory level.

The importance of smoking as the source of pleasant sensations of taste and smell is difficult to judge. Anecdotal evidence suggests these factors are more likely to play a role in the enjoyment of pipe and cigar smoking than cigarette smoking.[13] However, cigarette advertisements stress the importance of taste and flavour of particular brands and panels of trained testers are widely used in the industry. Tar is believed to be the major determinant of taste and flavour and there is some evidence for at least partial compensation of tar. Sutton *et al.*[14] showed that total puff volume was a better determinant of plasma nicotine than the tar or nicotine yield of the cigarette, its length or the reported number of cigarettes smoked on the test day. When nicotine yield was held constant, low tar smokers increased their puff volume and had higher plasma nicotine levels compared to higher tar smokers suggesting they were compensating for a reduced delivery of tar. Data from carboxyhaemoglobin analyses were consistent with this interpretation. The findings of an early study by Goldfarb,[15] which independently varied the tar and nicotine yields of experimental brands of cigarettes (though over a relatively narrow range), did not support those of Sutton. However, a role, if any of tar should not be dismissed on the basis of these two studies alone.

Of the other major smoke constituent, carbon monoxide, there appears to be no evidence to implicate it in compensatory smoking. Indeed, as mentioned previously, the published information on other possible factors associated with compensatory smoking is extremely sparse. It is quite likely that studies have been undertaken, but the results are not publicly available.

15.6. Experimental problems

Problems in assessing which factor or combination of factors a smoker compensates for are vast, but fall into two distinct groups. In the first, selection of the most appropriate method for estimating compensation is now the subject of intense debate. Since the variables in human smoking behaviour (e.g. number of cigarettes smoked, volume of smoke drawn per puff, puff duration, smoke manipulation in the mouth and nose, amount of smoke wasted prior to inhalation, amount of diluting air, degree/depth of inhalation, retention time of smoke in the lungs) all contribute towards compensation it becomes difficult to compare results obtained in the numerous studies when different endpoints are assessed.[7,16] The second group of problems relates to the examination of the relative contribution to smoking behaviour from nicotine, tar, carbon monoxide and sensory factors. It is important that these can be varied independently from one another but the technical difficulties encountered in designing cigarettes to meet these criteria for human studies are clearly very great, especially when taste and flavour are to be held constant. Some studies have attempted to manipulate tar and nicotine deliveries independently, and these suggest that the smoker's response to a given cigarette cannot be explained on the basis of nicotine yield alone.[17] Indeed, this latter point may possibly explain why so little attention has been focused on other possible factors since the evidence for nicotine titration is so conflicting.

15.7. Conclusions

It is clear that nicotine is not the only factor which determines smoking behaviour; other factors such as cigarette design are also of paramount importance. Although the influence that taste, flavour, and common chemical sense stimulation have on smoking behaviour has been known for a long time, only recently have techniques become available for research into this important area. It is anticipated that knowledge will expand rapidly. Whilst it may be true to say that no experiment has clearly demonstrated that 'non-nicotine' factors are of major importance to the smoker, it would also appear true to say that no study has shown unequivocally that nicotine alone determines smoking behaviour.

Acknowledgements

The author would like to acknowledge the help and advice given by Mrs Cheryl Swann of the Tobacco Products Research Trust in the preparation of this manuscript.

References

1. Tso, T. C. (1980). Modification through agricultural techniques for developing a safer tobacco. In *Banbury Report 3—A safe cigarette?* (ed. G. Gori and F. Bock), pp. 181–90. Cold Spring Harbor Laboratory, New York.
2. Selke, W. A. (1980). Reconstituted tobacco sheet. In *Banbury Report 3—A safe cigarette?* (ed. G. Gori and F. Bock). pp. 205–13. Cold Spring Harbor Laboratory, New York.
3. Eicher, T., Muller, F., and Steinhoff, D. (1980). Improving cigarettes with a cotobacco material. In *Banbury Report 3—A safe cigarette?* (ed. G. Gori and F. Bock), pp. 215–24. Cold Spring Harbor Laboratory, New York.

4. Keith, C. H. (1980). Physical methods for modification of tobacco smoke. In *Banbury Report 3—A safe cigarette?* (ed. G. Gori and F. Bock), pp. 225–37. Cold Spring Harbor Laboratory, New York.
5. Lavoie, E. J., Hecht, S. S., Hoffmann, D., and Wynder, E. L. (1980). The less harmful cigarette and tobacco smoke flavours. In *Banbury Report 3—A safe cigarette?* (ed. G. Gori and F. Bock), pp. 251–60. Cold Spring Harbor Laboratory, New York.
6. Gori, G. B. (1980). Less hazardous cigarettes: theory and practice. In *Banbury Report 3—A safe cigarette?* (ed. G. Gori and F. Bock), pp. 261–79. Cold Spring Harbor Laboratory, New York.
7. Rawbone, R., Murphy, K., Tate, M. E., and Kane, S. J. (1978). The analysis of smoking parameters. In *Smoking behaviour* (ed. R. E. Thornton), pp. 171–94. Churchill Livingstone, London.
8. Rawbone, R. (1984). The act of smoking. In *Smoking and the lung* (ed. G. Cumming and G. Bonsignore), pp. 77–93. Plenum Press, London.
9. Dunn, P. J. (1978). The effects of a reduced draw resistance cigarette on human smoking parameters and alveolar carbon monoxide levels. In *Smoking behaviour* (ed. R. E. Thornton), pp. 203–7. Churchill Livingstone, London.
10. Cain, W. S. (1980). Sensory attributes of cigarette smoking. In *Banbury Report 3—A safe cigarette?* (ed. G. Gori and F. Bock), pp. 239–49. Cold Spring Harbor Laboratory, New York.
11. Rose, J. E., Tashkin, D. P., Ertle, A., Zinser, M., and Lafer, R. (1985). Sensory blockade of smoking satisfaction. *Pharmacology and Biochemistry of Behaviour*, **23**, 289–93.
12. Russell, M. A. H. and Feyerabend, C. (1970). Cigarette smoking: a dependence on high nicotine boli. *Drug Metabolism Reviews*, **8**, 29–57.
13. Ashton, H. E. and Stepney, R. (1982). The importance of nicotine. In *Smoking psychology and pharmacology*, pp. 18–29. Tavistock, London.
14. Sutton, S. R., Russell, M. A. H., Iyer, R., Feyerabend, C., and Saloojee, Y. (1982). Relationship between cigarette yields, puffing patterns and smoke intake; evidence for tar compensation. *British Medical Journal*, **285**, 600–3.
15. Goldfarb, T., Gritz, E. R., Jarvik, M. E., and Stolerman, I. P. (1976). Reactions to cigarettes as a function of nicotine and tar. *Clinical Pharmacology and Therapeutics*, **19**, 767–72.
16. Nil, R., Buzzi, R., and Battig, K. (1986). Effects of different cigarette smoke yields on puffing and inhalation: is the measurement of inhalation volumes relevant for smoke absorption? *Pharmacology Biochemistry and Behaviour*, **24**, 587–95.
17. Ashton, H. E. and Stepney, R. (1982). Self regulation of nicotine intake. In *Smoking psychology and pharmacology*, pp. 66–90. Tavistock, London.

16

Factors influencing choice of low-tar cigarettes

MARTIN JARVIS, ALAN MARSH, and JIL MATHESON

Abstract

There has been little explicit study of the determinants of brand choice in smokers, but survey data consistently point to differences between low-tar and other smokers. Low-tar smokers are more likely than smokers of higher yielding brands to be older, female, and middle-class. Their nicotine intakes appear to be similar to, or slightly less than, those of other smokers. In terms of attitudes to smoking, they express greater motivation to give up, stronger intention to try, and are more confident of succeeding. These observations are consistent with the idea that for some smokers the switch to low-tar brands may be a preliminary move towards giving up altogether.

16.1. Introduction

Low-tar cigarettes were introduced to the British market in the early 1970s. Their definition is in terms of tar yield (up to 1984 those containing less than 11 mg tar; since 1985, less than 10 mg), but they tend also to be low in nicotine and carbon monoxide delivery by comparison with other cigarettes. The UK health authorities have consistently recommended that those smokers who are unwilling or unable to give up the habit completely should switch to lower yielding brands, on the assumption that this might reduce health risks from smoking. Despite this, low-tar brands have appealed to only a relatively small minority of smokers. The present paper is concerned not with whether low-tar smoking is in fact associated with any lowering of risk, but with the factors that lead smokers to opt for these brands. In particular, an attempt will be made to answer the following questions: Is it mainly lighter smokers who are attracted to low-tar cigarettes? Does the availability of low-tar cigarettes encourage the uptake of smoking by young people? Do people switch to lower yielding brands through a concern to minimise risks to their health? Can switching to low-tar cigarettes be construed as a preliminary move towards giving up smoking altogether?

16.2. Prevalence of low-tar smoking and demographic profile

Low-tar cigarettes were smoked by less than 1 per cent of cigarette smokers in 1972. Their market share expanded quickly from this low base to reach about 15 per cent in 1977 (see Fig. 16.1). Since then there has been little further change, and in 1985 they were smoked by only 17 per cent of British smokers of manufactured cigarettes.[1,2] The extent to which low-tar smoking is a minority activity is further documented by data from the 1984 General Household Survey (Table 16.1).[3] Among women, smoking can largely be equated with the smoking of manufactured cigarettes, but almost a third of male smokers smoke tobacco products other than manufactured cigarettes (own-rolled cigarettes, cigars, etc.). As a result, 11 per cent of male smokers of manufactured cigarettes were low-tar smokers in 1984, but only 8 per cent of male smokers overall. The corresponding figures for women were 23 and 22 per cent, respectively.

The distinctive demographic profile of the low-tar smoker has been established by numerous surveys. Compared with smokers of higher-yielding brands, low-tar smokers are more likely to be older, to be female, and to belong to a higher socioeconomic group. They are also likely to smoke slightly fewer cigarettes per day. Table 16.2 shows the pattern of low-tar

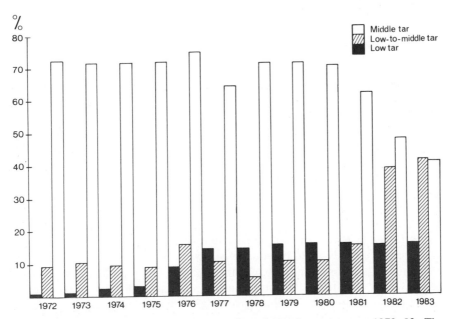

Fig. 16.1. United Kingdom cigarette market shares by tar group 1972–83. The market share of low-tar cigarettes has stagnated since 1977. See Jarvis and Russell.[1] for method of estimation.

Table 16.1. Product smoked by sex: Great
Britain 1984 (%)

	Men	Women
Manufactured cigarettes	30	32
Of whom: Low-tar	11	22
Non-low-tar	83	72
Unknown	6	6
Own-rolled cigarettes	6	1
Cigars	6	0
Any Smoking	43	33
Base	8377	9681

Note: 0 indicates less than 0.5 per cent.
Source: General Household Survey.[3]

smoking from the November 1985 survey carried out by National Opinion
Polls on behalf of the Office of Population Censuses and Surveys.[4] An
estimated 24 per cent of women smokers of manufactured cigarettes were
smoking low-tar brands, compared with 10 per cent of men; and 32 per
cent of professional and managerial cigarette smokers smoked low-tar
brands as against only 14 per cent among semi-skilled and unskilled manual
workers. Eight per cent of smokers aged 16–24 smoked a low-tar brand
compared with 25 per cent of those aged 55 or above.

16.3.   Do lighter smokers smoke low-tar brands?

Several observational studies have compared measures of smoke intake in
low-tar (or ventilated filter) cigarette smokers with those smoking higher
yielding unventilated filter cigarettes. Russell *et al.*[5] found no significant
difference in blood nicotine or COHb concentrations, or in cigarette consumption, indicating a high degree of compensatory inhalation for the
reduction in cigarette yields. The importance and extent of smoker compensation was confirmed by Wald[6] in a study of 2455 men attending for a
health screening examination. Smokers of ventilated and unventilated
cigarettes were found to compensate almost completely for the reduction
in carbon monoxide yield by comparison with smokers of plain cigarettes.
More recently, Russell *et al.*[7] reported statistically significant reductions
of 11–17 per cent in COHb, and plasma nicotine and cotinine concentrations in low-tar smokers, while cigarette consumption was closely similar to
that of smokers of other brands.

These studies agree in showing at most a minor reduction of intake in
low-tar smokers. Most low tar smokers are people who have previously

Table 16.2. Low-tar smoking by age, sex, and class: Great Britain November 1985 (%)

| | Base | Age | | | | | | Class | | | | | |
		16–24	25–34	35–44	45–54	55–64	65 +	AB	C1	C2	DE	Total
Men	406	3	6	17	9	13	23	21	13	6	10	10
Women	478	15	15	24	30	25	39	43	32	18	18	24
All smokers	884	8	11	21	22	20	32	32	24	12	14	18

Base: smokers of manufactured cigarettes with known brand preference.
Note: 16 per cent of male cigarette smokers and 2 per cent of female preferred own-rolled cigarettes and are therefore excluded.
Source: National Opinion Polls.[4]

smoked higher yielding brands. There are two potential explanations for the slight reduction in smoke intake seen in low-tar smokers. The first is that these people were somewhat lighter smokers before they switched down and, after switching, they compensated so as to maintain their previous intake. The second is that compensation was less than complete when they switched so that the low-tar cigarettes did lead to a small reduction in exposure. The data presently available do not permit a choice between these alternatives. What does seem clear, though, is that any differences between low-tar and other smokers in intake of nicotine or carbon monoxide are minimal. It therefore seems unlikely that a major role in choosing to smoke low-tar cigarettes can be attributed to being a lighter, less dependent smoker.

16.4. Low-tar smoking and novice smokers

Concern has sometimes been expressed that the availability of lower-yielding milder brands may contribute to the uptake of smoking by young people by making inhalation easier and less irritant. As shown above, low-tar smoking is least common in young adults and becomes steadily more popular in older age groups. While these data do not seem to support the notion that young people start on low-yielding brands and switch to stronger cigarettes as their habit develops, there remains a possibility that some such effect may be found in younger teenage smokers below the age of 16. There are at present only limited data available on brand preference among schoolchildren who smoke. In a survey in Manchester, Ledwith found that the middle tar brands which were most popular with adults also dominated the preferences of children.[8] Similar preferences were reported by McNeill et al.[9] Among their sample of 597 schoolgirls from an inner London comprehensive, 91 per cent of those who smoked had a preferred brand, but only 10 per cent of these preferred low-tar brands. These girls have now been followed for a further year, and brand preferences in both 1985 and 1986 have been established. While the great majority of girls smoked middle-tar brands in both years, there was some brand switching, with switching to low-tar cigarettes from higher-yielding brands being significantly more frequent than switches in the other direction (A. D. McNeill and M. J. Jarvis, unpublished observations). These data have obvious limitations in being based on a small sample of smokers drawn from a single school. Nevertheless, they do not suggest that low-tar cigarettes play an important role in establishing the smoking habit in teenagers. Rather they agree with other data in suggesting that an initial preference for middle tar brands is followed by switching to low-tar with increasing age and smoking experience. Further longitudinal studies of brand preference in larger and more representative samples are needed.

16.5. Smoking history, smoking attitudes, and low-tar smoking

The amount of information available on low-tar smokers' attitudes towards smoking and its effects, and towards giving up smoking is scanty. Similarly, we know little about the smoking history which leads people to a switch to low-tar cigarettes. The large-scale survey of smoking attitudes and behaviour carried out by the Office of Population Censuses and Surveys in 1981[10] appears to be unique in having examined attitudinal variables in smokers of brands in the different tar groups. In published analyses it has been shown that low-tar smokers were more likely than other smokers to think that smoking low-yield cigarettes reduces the chance of getting smoking related diseases (63 *v.* 37 per cent), whilst those who smoke higher yielding cigarettes were more likely to feel that cigarette yield makes no difference to risk (57 *v.* 32 per cent among low-tar smokers). This might indicate that those who switch to low-yield cigarettes do so from a belief that this will reduce their risk. However, an alternative explanation might be that expressed attitudes make sense of the person's own behaviour. In other words, attitudes may be congruent with chosen behaviour, but need not necessarily have played a crucial role in the adoption of that behaviour.

Here we present further analyses of the OPCS data in order to explore possible differences between low-tar and other smokers in smoking history, subjective dependence on smoking, desire and resolve to stop, and confidence in succeeding in a cessation attempt. These analyses compare 552 low-tar smokers with 1728 smokers of higher yielding brands. Full details of the study design and questionnaire are given elsewhere.[10]

Table 16.3 shows the mean values for low-tar and other smokers on demographic variables and cigarette consumption. As in other surveys, those who smoked low-tar brands tended to be older, female, to belong to a higher socio-economic group, and to smoke somewhat fewer cigarettes

Table 16.3. Characteristics of low-tar and other smokers: OPCS survey of smoking attitudes and behaviour

	Low-tar ($n = 552$)	Others ($n = 1728$)
Age	43.9	35.9
% Female	30.3	16.7
% Non-manual	30.3	18.5
Daily consumption	17.4	19.0

Base: smokers of manufactured cigarettes with known brand.
Source: Marsh and Matheson.[10]

per day. In a multiple regression analysis each of these variables had a significant independent association (*P*<0.001) with low-tar smoking. Analysis of covariance was therefore used to control for these variables in assessing possible differences between low-tar and other smokers on attitudinal and other questionnaire items.

Table 16.4 gives comparisons between low-tar and other smokers in summary form. Low-tar smokers said that they had taken up smoking marginally later than others, and a higher proportion of them (76 *v.* 68 per cent) reported having made an attempt to give up smoking altogether. They had had more cessation attempts which had lasted longer than a week, and their longest time off smoking was for about 4–5 months compared with just under 3 months in smokers of higher yielding brands. Although some questions designed to tap subjective aspects of dependence

Table 16.4. Smoking attitudes and behaviour in low-tar and other smokers

	Low-tar (*n* = 552)	Others (*n* = 1728)	significance
Smoking history			
Age started smoking (years)	17.4	17.1	0.06
Ever tried to give up altogether (%)	76	68	***
No. times given up for more than a week	1.98	1.74	**
Duration of longest abstinence (weeks)	19	11	***
Dependence			
Need for cigarettes	2.93	2.85	NS
How often smoke without enjoying it	2.06	2.05	NS
Difficulty not smoking for half a day	1.46	1.49	NS
whole day	2.01	2.12	**
week	2.62	2.79	***
(1 = not at all difficult ... 4 = impossible)			
Desire to give up (1 = very much ... 7 = not at all)	3.37	3.71	**
Resolve to try (1 = will definitely try ... 7 = will never try)	3.31	3.59	**
Confidence in succeeding (1 = sure I would succeed ... 7 = sure I would fail)	3.42	3.72	**

**P < 0.01, ***P < 0.001.

Note: adjusted mean values are shown after allowing for the effects of the covariates age, sex, socioeconomic group, and cigarette consumption.

(e.g. 'Some smokers say that they will do almost anything to get cigarettes when they run out. How true is this for you?') showed no significant differences, low-tar smokers rated stopping for a day or for a week less difficult than did others. There was, therefore, some indication, realistic perhaps in view of their smoking history, that low-tar smokers saw smoking as a somewhat easier habit to break.

Attitudes towards giving up were examined by three questions: 'How much do you *want* to give up smoking?' (desire); 'How likely is it that you *will try* to give up smoking?' (resolve); 'And if you tried, how likely is it that you would succeed in giving up smoking?' (confidence). There were differences between low-tar and other smokers in their replies to each of these. Their desire to give up was greater, their resolve to try was stronger, and they were more confident of succeeding. However, although these differences were highly significant statistically, their magnitude was not great, averaging less than one-half of one scale point in each case.

16.6. Discussion

Low-tar cigarettes have been a feature of the British market for over a decade, but there has been little systematic study of who smokes them and why. At present, therefore, we have no thorough understanding of factors associated with their use. Certain hypotheses are not supported by the evidence available, but other more or less plausible ideas can be advanced for further study.

Smokers of low-tar cigarettes do not appear to be especially light smokers who are satisfied by much lower smoke intakes than others. Nor does it seem to be the case that low-tar cigarettes serve to provide an easy introduction to smoking for those teenagers who are in the early stages of the habit. The greater prevalence of low-tar smoking among women is intriguing, but may do no more than reflect the fact that the dominant low-tar brand in Britain is currently promoted by its manufacturer with a 'feminine' image. In other countries the advertising images associated with low-yield cigarettes may involve sport and youth, with consequent differences in the demographic profile of those who smoke them (see Rickert, chapter 7). Against this, the demographic profile of the low-tar smoker in the United States is similar to that found in Britain.[11]

There are a number of indications that low-tar smokers are people who are relatively more concerned about reducing health risks and about giving up smoking altogether. They are older, and giving up smoking bears a strong relationship to age. They are also drawn more from the higher socio-economic groups which respond more readily to health information and have a lower smoking prevalence. On questions tapping desire to give up, intention to try, and confidence in succeeding they consistently score as more concerned than other smokers, even when potential confounders such as age, sex, class, and cigarette consumption are controlled for.

In conclusion, the following can be offered as a tentative characterization of low-tar smoking. People switch to low-tar cigarettes as they progress in their smoking careers and become more concerned about health risks. They believe (or hope) that smoking low-tar cigarettes may be less hazardous, so the switch can be seen as a way station along the road to giving up altogether. Indeed, low-tar smokers are more motivated to give up and more confident of succeeding, so that it seems probable that they will, in fact, be somewhat more likely to become ex-smokers than those who smoke higher yielding brands. Perhaps low-tar smoking should be seen as a preliminary move towards non-smoking by the relatively better informed and more concerned (though not necessarily less dependent) smoker.

Acknowledgements

Financial support was provided by the Medical Research Council.

References

1. Jarvis, M. J. and Russell, M. A. H. (1985). Tar and nicotine yields of UK cigarettes 1972–1983: sales-weighted estimates from non-industry sources. *British Journal of Addiction*, **80,** 429–34.
2. Jarvis, M. J. and Russell, M. A. H. (1986). Data Note 4. Sales-weighted tar, nicotine and carbon monoxide yields of UK cigarettes 1985. *British Journal of Addiction*, **81,** 579–81.
3. Office of Population Censuses and Surveys (1986). General Household Survey 1984. London, HMSO.
4. National Opinion Polls (1985). A report on a survey carried out by NOP Market Research Limited for Office of Population Censuses and Surveys. November 1985.
5. Russell, M. A. H., Jarvis, M. J., Iyer, R., and Feyerabend, C. (1980). Relation of nicotine yield of cigarettes to blood nicotine concentrations in smokers. *British Medical Journal*, **280,** 972–6.
6. Wald, N. J., Boreham, J., and Bailey, A. (1984). Relative intakes of tar, nicotine and carbon monoxide from cigarettes of different yields. *Thorax*, **39,** 361–4.
7. Russell, M. A. H., Jarvis, M. J., Feyerabend, C., and Saloojee, Y. (1986). Reduction of tar, nicotine and carbon monoxide intake in low tar smokers. *Journal of Epidemiology and Community Health*, **40,** 80–5.
8. Ledwith, F. (1984). Does tobacco sports sponsorship on television act as advertising to children? *Health Education Journal*, **43,** 85–8.
9. McNeill, A. D., Jarvis, M. J., and West, R. J. (1985). Brand preferences among schoolchildren who smoke. *Lancet*, **ii,** 271–2.
10. Marsh, A. and Matheson, J. (1983). *Smoking attitudes and behaviour*. HMSO, London.
11. Stellman, S. D. and Garfinkel, L. (1986). Smoking habits and tar levels in a new American Cancer Society prospective study of 1.2 million men and women. *Journal of the National Cancer Institute*, **76,** 1057–63.

17

The role of nicotine in the tar reduction programme

SIR PETER FROGGATT and NICHOLAS J. WALD

Abstract

In the smoking habit nicotine has both advantages and disadvantages. Nicotine contributes to the apparent pleasure of smoking and it improves performance in smokers but an important disadvantage is the dependence it induces. Long-standing speculation that nicotine may be a cause of cardiovascular disease does not seem well-founded. Nicotine may be a co-carcinogen through its possible role in the formation of nitrosamines though the view that it is the nicotine in tobacco smoke that is largely responsible for the cancer induced by smoking is not supported by the epidemiological evidence. It is possible that nicotine may contribute to the anti-oestrogenic effect of smoking, though this is uncertain and requires further research. Nicotine, while not the only factor in controlling overall smoking behaviour, is recognized as being an important factor in regulating compensatory smoking. This can be used to advantage by reducing nicotine yields less than tar yields and thereby reducing the intake of tar as yields are gradually reduced. Reduction in tar yields will reduce the incidence of lung cancer and will probably also reduce the incidence of chronic obstructive lung disease, but there is little evidence that it will affect the incidence of ischaemic heart disease. Trend data on smoking and its related diseases support these conclusions. On present evidence nicotine yields should be brought down, but it would seem that the toxicity of cigarettes may be reduced more if nicotine yields are reduced to a lesser extent than tar yields.

To the smoker nicotine confers undoubted benefits though it also has disadvantages. The benefits are well documented. There is no doubt that nicotine improves performance in smokers. It may also do so in non-smokers as evidenced, for example, by the observation cited by Jarvis in Chapter 12 that nicotine administered nasally improves non-smokers' performance on a simple task of psychomotor speed, the enhancement being blocked by mecamylamine, a nicotine antagonist. Nicotine relieves stress, enhances mood, and improves concentration at least in smokers. Further-more, the evidence shows that nicotine plays an important role in in-fluencing 'compensatory smoking'. It follows that by maintaining the nico-tine yields of cigarettes while *pari passu* lowering the tar yield, tar intake can be reduced more than by lowering nicotine and tar in equal propor-tions. Understanding the determinants of compensatory smoking in this way is important in judging the health effects of changing cigarette yields. While compensation occurs it is reassuring to know that *it is not complete*, thus implying that while tar reduction may not produce its benefits to the extent which might initially have been supposed, it *does*, nonetheless, lead to a reduced tar intake.

What are the disadvantages of nicotine? One is the dependence it produces in smokers and this can be considerable; tobacco smoking is still the most popular mode of nicotine delivery. The dependence on nicotine is a two-edged sword since the fact that people smoke, in part, for nicotine can be argued both as a reason to *maintain* nicotine in a tar reduction programme (to reduce compensation), and to *lower* it in order to wean the smoking population off the habit or discourage the development of dependence in new smokers—a particularly relevant point with the young. Certainly, this latter point argues that nicotine yields be not *increased*. The balance of the argument is, we believe, on present knowledge in favour of the view that nicotine yields be reduced less drastically than tar yields.

From the toxicological point of view nicotine has been mainly considered in relation to two groups of disorders, namely cardiovascular disease and, more recently, cancer. That nicotine has a role in the cause of cardiovascu-lar disease has its adherents, but the evidence is not compelling. The fact that pipe smokers, who have high intakes of nicotine, do not have a materially increased risk of ischaemic heart disease, on the face of it dismisses chronic exposure to nicotine as a significant cardiovascular hazard. The possibility that acute nicotine exposure (by, say, 'pulse-dosing' through deep inhaling) might be hazardous is likewise largely dismissed by the observation that smokers who have fatal heart attacks do not have them disproportionately after the first or second puff from the first cigar-ette of the day. Moreover, nicotine is not an agent which can induce chronic arterial damage in animals.

A major focus of attention in these Proceedings is the possibility that nicotine may act as a co-carcinogen. Nicotine itself is not genotoxic. There is no laboratory evidence that it is a carcinogen or that it enhances the activity of known carcinogens. On the other hand, nicotine alkaloids present in tobacco can be nitrosated to produce nitrosamines for which there is conclusive evidence of carcinogenicity in animals.

Nitrosamines are produced when secondary and tertiary amines are nitrosated. Nicotine is a tertiary amine and the tobacco alkaloids nornicotine, anabasine, and anatabine are secondary amines. Nitrosation in tobacco results from reaction with nitrite and nitrogen dioxide which is derived from the nitrate in the tobacco which, in turn, is associated with the use of nitrogenous fertilizers. The nitrate is present in the greatest quantities in the roots and stems of the tobacco leaf and the nitrosamines arise during tobacco curing and processing. In the tobacco smoke about one-third of nitrosamines come directly from the tobacco while perhaps up to two-thirds are synthesized during the burning of the tobacco (pyrolysis).* It is likely, therefore, that factors that lead to an increase in the nitrate concentration of tobacco, or in the concentration of secondary and tertiary amines, or in the yield of nitrogen dioxide will all lead to an increase in the total delivery of tobacco specific nitrosamines. There is good evidence that the nitrate content of tobacco affects the levels of nitrosamines in tobacco products and in smoke. There is, however, less evidence that the nicotine yield of a cigarette has a similar effect although Hoffmann in his paper presented data to show that the addition of nicotine to a cigarette can increase its yield of N'-nitrosonornicotine (NNN), one of the principal tobacco-specific nitrosamines. The relationship between the nicotine yield of a cigarette and its *total* yield of tobacco-specific nitrosamines is still unclear, though nitrosamine delivery appears to be much more influenced by nitrate yield than by nicotine yield.

On general grounds, we must accept that the increase in a secondary amine, such as nicotine, in tobacco is likely to lead to an increase in the yield of total nitrosamines, though the magnitude of such an effect is likely to be small in comparison with other factors influencing the yield of nitrosamines. Thus, US and French cigarettes have much higher nitrosamine yields than British cigarettes. US tobacco has a high nitrate content and French cigarettes have high yields of nitrogen oxides. A typical US filter cigarette will yield about 150 ng of NNN, 200 ng of NNK [4-(N-methyl-N-nitrosamino)-1-(3-pyridyl)-1-butanone], and 150 ng NAT (N'-nitrosoanatabine); and French cigarettes have similar yields. A typical UK filter cigarette has yields only about one-tenth of these levels while a small cigar yields nearly 10 times more. The yield of nicotine could not account for the 10-fold variation in nitrosamine delivery between, say, US and UK cigarettes or the 10-fold difference between the yield from small cigars and US cigarettes. Also, if the relative toxicity of US cigarettes were 10 times greater than that of UK cigarettes why is the risk of lung cancer in the US and Britain similar among men of the same age who smoke the same amount, and who started to smoke at the same age? Unquestionably, it would be wise to engineer cigarettes to reduce nitrosamine delivery, but

*It is also possible that some nitrosamine can come from metabolism in the body of absorbed nicotine, though this has not been firmly demonstrated.

altering the nicotine yield is not the best way to accomplish this. In any case, we must not forget that whatever toxic effects tobacco-specific nitrosamines may have, the carcinogenicity of tobacco smoke cannot be satisfactorily explained in terms of its nitrosamine delivery.

It is possible, though not yet proven, that nicotine may contribute to the anti-oestrogenic effect of tobacco smoking and the consequential effect on the risk of oestrogen-related disorders among smokers. Clarifying the role of nicotine is important since it may represent the most serious toxic effect of nicotine in humans. At present, there is too little information upon which to make a judgement and the position must be re-examined when further data are available.

We may therefore say that, at present, there are no strong reasons for believing that nicotine is an important *toxic* component of tobacco smoke though we cannot completely exclude the possibility that it may play a role in co-carcinogenesis and in the development of hormone-related disease.

Nicotine, though an important factor in regulating compensatory smoking and therefore a determinant of tar intake, is not the only factor. It is clear from the presentations at the Symposium that compensation is not due to a single mechanism and, indeed, nicotine does not determine all aspects of smoking behaviour as evidenced by the large number of smokers who manage to give up smoking altogether and by the observation that in certain countries, such as West Germany, cigarettes with relatively low nicotine yields are widely used. It is possible that people starting for the first time to smoke may be satisfied by a relatively low nicotine delivery cigarette, whereas established smokers who have got used to a higher nicotine delivery cigarette could not easily or immediately accept less nicotine.

Table 17.1. Trends in lung cancer mortality ratios by sex and age 1950–84 (England and Wales)

Sex	Year	Mortality ratios for age group				
		30–34	40–44	50–54	60–64	70–74
Men	1950–54	1.00	1.00	1.00	1.00	1.00
	1960–64	0.97	0.91	1.03	1.52	2.04
	1970–74	0.65	0.75	0.89	1.49	2.88
	1980–84	0.35	0.48	0.66	1.26	2.72
Women	1950–54	1.00	1.00	1.00	1.00	1.00
	1960–64	0.73	1.33	1.49	1.49	1.40
	1970–74	0.53	1.35	2.33	2.34	2.26
	1980–84	0.47	1.13	2.35	3.45	3.56

Table 17.2. Trends in weekly consumption of manufactured cigarettes per person 1950–84 (Great Britain, sales adjusted)

Sex	Year	Age		
		30–34	35–49	60 +
Men	1950–54	83	84**	39
	1960–64	79	87	51
	1970–74	84	87	50
	1980–84	56*	56	32***
Women	1950–54	36	26**	9
	1960–64	42	46	13
	1970–74	60	59	18
	1980–84	46*	51	17***

*25–34; **35–59; ***65 +.

The data presented in Chapter 5 on smoking trends are encouraging, especially those in men and young women. The major concern is the persistence of relatively high rates of cigarette smoking among teenage girls and persons in the lower socio-economic groups. The decline in the consumption of cigarettes, both expressed on a per person basis and as a percentage of smokers in the general population, shows that a low tar smoking policy can be successfully implemented *pari passu* with a policy designed to reduce smoking in general. This is important because trends in smoking-related diseases support the view, principally based on the results of epidemiological studies, that the reduction in tar yields is associated with the reduction in lung cancer. This effect has been most marked in the younger age groups in which exposure to high tar cigarettes in the past would have been less than in older smokers.

Tables 17.1–17.3 show how lung cancer mortality in men aged 30–34 years (an age group young enough to be less affected by past smoking habits than older men) in 1980–84 were only one-third of the mortality 30

Table 17.3. Trends in annual sales-weighted tar yield (mg/cigarette) (United Kingdom)

Year	Annual sales weighted tar yield (mg/cig)
1948–54	30
1962–68	26
1970–74	21
1980–84	16

years before, while cigarette consumption declined to two-thirds and tar yields to nearly one-half.* The decline in lung cancer can not be satisfactorily explained by the fall in cigarette consumption alone, but it can by the fall in tar-weighted cigarette consumption. Compensatory smoking might have been expected to attenuate the fall in tar yields, but there are grounds for believing that increases in the extent of inhaling probably leads to relatively less deposition of smoke particles on the proximal bronchial airways and, hence, less lung cancer than might otherwise be expected (though not necessarily less than would be expected with diseases that may be associated with the contact or absorption of smoke particles in the peripheral parts of the lung).

The association between tar-weighted cigarette consumption and lung cancer in young British men provides important support for the view that the reduction in tar yields have been beneficial. The reduction in lung cancer in young *women* supports this thesis, but the reduction appears greater than would be expected from the decline in tar-weighted cigarette consumption—lung cancer fell to about half over the same 30-year period, while tar-weighted cigarette consumption declined to about two-thirds (28 per cent increase in consumption and a 47 per cent decrease in tar yields). While the reason for this striking reduction in lung cancer is not clear it may have arisen because the decline in tar yields in young women may have been greater than the average reduction for the population as a whole. It may also have been due to other environmental changes, notably improvements in the control of air pollution in Britain during the 1950s. The percentage reduction in lung cancer that may have arisen as a result of cleaner air is likely to be more apparent in young women with their low background rates than in men whose rates were much higher.

There is also evidence, though less compelling, that reducing tar yields has reduced the risk of chronic obstructive lung disease. On the other hand, the data on ischaemic heart disease and aortic aneurysm are equivocal. However, on the basis of epidemiological studies (not cited in these Proceedings) and on the basis of the trends (that are described), there is no evidence that the prevalences of these diseases have been detrimentally affected by decreases in tar yields. Thus, the conclusion *viz*, that the product modification programme has resulted in net benefits, is sound and supported by the evidence.

It is, of course, possible that factors other than tar yields and the prevalence of cigarette smoking have contributed to the reduction in lung cancer, for example a reduction in the specific carcinogenicity of tobacco smoke. The accompanying table, however, which shows declines in lung cancer in younger age groups over some 30 years reflecting declines in tar yield,

*This assumes that the secular change in tar yields affected all age groups to the same extent; it probably had a greater impact on the young than the old.

suggests that it is not necessary to invoke such an explanation though, clearly, it cannot be excluded solely from the trend data.

Several speakers during the Symposium expressed the view that many smokers who switch to low tar cigarettes do so preparatory to stopping smoking altogether, and that this was not simply an intermediate step in a self-selected regimen towards giving up the habit, but could in itself actively help the smoker who wants to stop smoking achieve this goal. If these views are substantiated then a more active policy of inducing smokers onto low tar brands would be justified; for the last five years the proportion of smokers smoking low tar brands in the UK has plateaued at 12–15 per cent. The low tar policy would then be a useful step, not only in reducing the hazard to continuing smokers, but also as a way of converting such smokers into ex-smokers.

Index